S0-DUT-748

Fungi, man and his environment

R. C. Cooke

Fungi, man and his environment

Longman
London and New York

Longman Group Limited London

Associated companies, branches and representatives throughout the world

Published in the United States of America by Longman Inc., New York

First published 1977

Library of Congress Cataloging in Publication Data

Cooke, Roderic.
Fungi, man, and his environment.
Bibliography: p.
Includes index.
1. Fungi—Economic aspects. 2. Fungi. I. Title.
QK604.C633 589'.20'046 77-1460
ISBN 0-582-46034-4

Printed in Great Britain by
Butler & Tanner Ltd, Frome and London

Contents

Preface

An author who has much to communicate under this head, and expects to have it attended to, may be compared to a man who takes his friend by the button at a Theatre Door, and seeks to entertain him with a personal gossip before he goes in to the play.
Charles Dickens, Preface to *The Posthumous Papers of the Pickwick Club*, 1847

Many good textbooks concerned with various aspects of the study of fungi are available which it would be difficult to equal or to improve upon. The majority have been written by professional mycologists for the enlightenment of other professional mycologists, or as instruments for the training of students who aspire to be mycologists. By virtue of the audience to which they are addressed, they tend to present difficulties to the general biologist who wishes to know something of fungi, and to the naturalist or student who has a curiosity towards these organisms.

The progress of experimental mycology is firmly in the hands of predominantly laboratory-based research workers, usually specializing in taxonomic, physiological, biochemical, genetic or pathogenic aspects of fungal behaviour. These specialized activities are important, particularly where there are medical or agricultural problems to be solved, but in specializing there is a danger that the broad picture of fungi as whole organisms will be lost. It can quite fairly be said that a great many mycologists are concerned with the esoteric in their discipline, and that mycologists as a whole have a not undeserved reputation for unnecessary precision within their subject and for presenting it in a desiccated form. Yet fungi and their activities impinge on us all, literally daily, in a number of ways. They interact for good or bad with a vast range of other organisms and their intervention in human affairs in the widest sense is of great importance.

I am a professional mycologist and have for a number of years been concerned with a study of the arcana of my chosen fields of research. I am, however, fortunate in that I teach undergraduate and postgraduate students. In order to do this a broadness of approach has been useful especially when, within it, the uniqueness of fungi as organisms can be communicated. Drawing from this experience I have attempted to write a book about fungi rather than a textbook on fungi. Instead of attempting to cover the whole field of mycology I have selected mycological topics which I hope will illustrate what fungi do, why they are important to us, and how they are being studied at the present time.

I have tried to write both for those readers that have some mycological background and for those having little or none at all. To do this has not been easy, but I hope that I have for the most part reached a satisfactory compromise. In order to rapidly introduce readers with no knowledge of fungi to these organisms the first two chapters are concerned with what fungi are and how they differ from other living things. The first chapter in particular was written to provide a vocabulary of mycological terms and a simple outline of fungal behaviour, which allows subsequent chapters to be better understood. This chapter is primarily a brief phrasebook for a larger and infinitely more complicated mycological language. Much has been omitted and much simplified, and those with some knowledge of fungi may prefer to ignore this chapter. Throughout the remainder of the book complicated terminology has been reduced to a minimum and where the use of technical words has been unavoidable their meaning has been made as clear as possible.

No selection of topics can be fully satisfactory to everyone and mine might justifiably be called idiosyncratic. Many of those fields that I have ignored—for example, food microbiology, the production and action of antibiotics, and the role of fungicides—are by no means unimportant. Understanding them, however, requires some knowledge of biochemistry at a level which I feel is beyond the scope of this book. I can only trust that I have compensated for these shortcomings with the wide range of subjects that I have included. These are in the main those concerned with the direct interaction of fungi with plants or with animals, including man.

Having admitted to a degree of arbitrariness in the choice of material I must also confess that serendipity has played a part in determining some of the subject matter. When gaining background information on a particular topic I have sometimes been drawn, usually through a chain of obscure and oblique literature references, into an entirely different subject hitherto unknown to me. This is, of course, not a unique experience but I mention it here because it is one which many students quite purposely resist. Dismissed as mere browsing it is looked upon by them as a time-wasting activity with no concrete benefits. It can, however, result in an extension of personal knowledge within relatively unknown but not unimportant areas.

Some interactions between fungi and other organisms are of great economic importance and where possible I have tried to estimate this importance in terms of money or amounts of particular commodities. Unfortunately, meaningful figures are surprisingly difficult to obtain and, even where they are available, the rapid alteration of economic patterns which is a feature of the modern world undermines their accuracy. Figures presented here should therefore be accepted as a guide to the economic value of certain activities of fungi rather than as absolute data.

I have also sought to stimulate further reading, but access to the bulk of mycological literature is often difficult if the facilities of a university library are not available. This is because much information is to be found exclusively in specialized journals or monographs. The bibliography therefore includes some books and general scientific journals that can be obtained without too much difficulty with the aid of a good public library.

Finally, I must point out that I have in the main resisted the temptation to philosophize but have tried to give an account of some of the activities of fungi in such a way that they reveal themselves for what they have been to man in the past, what they are to us now and what they may be for us in the future.

I hope that despite the disparity of subject matter a broad theme does emerge throughout the book, namely, that fungi have always been used by and have fascinated man, and that many contemporary investigations into important, fundamental mycological problems have sprung directly from reappraisals of long-known aspects of fungal behaviour.

Roderic Cooke
Sheffield, November 1976

Acknowledgements

I am very grateful to Liz Craghill for her invaluable help with the line drawings and to Glyn Woods for preparing the photographs. I am also grateful to Janet Lambden for deciphering the first draft of the manuscript and to Nuala Ruttle who prepared the final copy.

The lines quoted at the beginning of Chapter Two are from the poem *Mushrooms* by Sylvia Plath which appears in *The Colossus* published by Faber & Faber, copyright Sylvia Plath 1967. I am indebted to Olwyn Hughes for permission to use this extract.

The following authors, publishers and editors kindly gave their permission for me to reproduce illustrations, sometimes in a modified form, from their books and journals:

Dr P. M. Austin Bourke, Dr C. M. Brasier, Dr M. Catt, Dr J. M. Cullen, Dr J. N. Gibbs, Dr P. F. Kable and the editor of *Nature*.

Longman Group Ltd, publishers of A. H. R. Buller's *Researches on Fungi*.

Dr E. Müller-Kögler and Paul Parey (Berlin and Hamburg), the author and publishers, respectively, of *Pilzkrankheiten bei Insekten*.

Dr J. Rishbeth and the editors of the *Annals of Applied Biology*.

Professor N. A. Weber and the American Philosophical Society, the author and publishers of *Gardening Ants–The Attines*.

The editors of *Science* (the journal of the American Association for the Advancement of Science).

John Wiley & Sons Inc., publishers of Paul Buchner's *Endosymbiosis of Animals with Plant Microorganisms*. (1965).

Dr O. L. Gilbert, Dr D. J. Read and the editors of *New Phytologist*.

Professor B. Lowy, Dr D. M. Macleod, Dr E. Müller-Kögler, Dr N. Wilding and the editor of *Mycologia*.

Mr C. Walker and the Forestry Commission.

Dr D. S. Hayman, Dr F. D. Podger, and Dr F. J. Newhook.

Annual Reviews Inc.

Cover illustration: Bruce Coleman Ltd; photographer, Jane Burton

List of illustrations

I

The new kingdom

No eye can see their beauties; their office is unknown; their varieties are not regarded; they are hardly allowed a place among Nature's lawful children, but are considered something abnormal, worthless and inexplicable.
W. D. Hay, *British Fungi*, 1887

It is now generally recognized that mycology, the branch of biology which is concerned with the investigation of all aspects of the life of fungi, constitutes a distinct discipline. This has not, however, always been the case and the emergence of mycology as a separate entity is relatively recent. Linnaeus, the eighteenth-century botanist and classifier of organisms, published a now famous aphorism, '*Lapides crescunt; vegetabilia crescunt & vivunt; animalia crescunt, vivunt & sentiunt*'. This reflected the long-held view that if a thing simply existed then it was mineral, if it lived then it was vegetable, and if it also had senses then it was animal. Using these criteria it was clear that fungi were not mineral and, since they had no discernable senses, they were obviously vegetable. The notion that all living things belonged to one of two great kingdoms, plants and animals, persisted into the second half of the present century (Fig. 1.1). Historically, mycology therefore inevitably grew as an integral but minor part of the study of botany. Fungi were certainly curious and interesting organisms but they were, however, often difficult to observe and were probably not particularly important. This long-standing association with botany is preserved even today in universities and other institutions where mycology is taught and where the subject is invariably presented as an aspect of plant sciences. There is no scientific reason why this should be so; for mycology is as far removed from botany as botany is from zoology, bacteriology or virology.

Modern systems of classification, based on a good understanding of evolutionary processes and on detailed information concerning the structure and function of organisms, have elevated the fungi to a high position among other life forms. This change has arisen fundamentally from considerations of how life may have begun on this planet. Interesting though they are, it is inappropriate to discuss in detail these considerations, or the arguments and hypotheses arising from them, except in as much as they affect the present status of fungi. It used to be thought that the first living things must have contained chlorophyll or other pigments which allowed them to use atmospheric carbon dioxide and sunlight to manufacture, through the process of photosynthesis, their major energy-rich nutrients. It was assumed that organic matter in quantity was lacking on the primitive earth, so that this was therefore the only way in which the first organisms could secure their own organic matter and incorporate more complex substances into their bodies. Life forms lacking photosynthetic pigments would have arisen later, being able to exploit the living or dead bodies of the pigmented organisms, and these unpigmented entities gradually evolved into animals. The pigmented organisms evolved into plants and then, at some point during this process, 'degenerate' forms appeared that had for some reason lost their chlorophyll and had thus become dependent for survival on organic matter already formed. These were the ancestors of the fungi and since fungi in many respects superficially resemble algae it was probable that they were derived from algae that had lost their pigments.

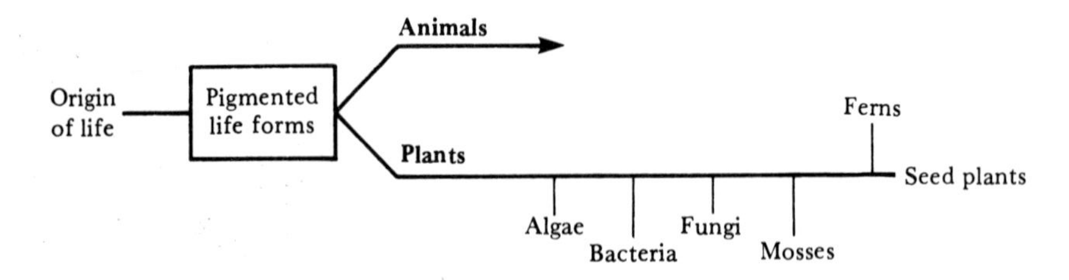

Fig. 1.1 A two-kingdom system for the classification of living things. The position of emergence of the sub-groups of Plants is arbitrary and does not represent their relative times of origin.

This view can no longer be held. It is now clear that life began under atmospheric conditions very different from those prevailing at present in which there were high levels of ammonia and methane together with a lack of oxygen. These conditions were favourable for the elaboration of organic compounds by direct chemical reaction without the necessity for photosynthetic processes. Life forms lacking chlorophyll, therefore, probably arose before those having chlorophyll, and the fungi are possibly derived from animal-like rather than plant-like ancestors.

Not only are fungi distinct from plants but they also have characteristics which set them apart from other major groups of living things. The old idea of the two kingdoms has now gone, being replaced by concepts which have led to the erection of at least four, and ideally five, kingdoms with the kingdom Fungi forming one of them. These kingdoms are based on the level of organization found in living things and on their mode of nutrition (Fig. 1.2). Three fundamental modes of nutrition exist: photosynthesis, in which organic matter is manufactured; absorption, in which soluble organic compounds are taken into the cell; and ingestion, in which solid organic particles are taken in. There are also a number of levels of organization ranging from a single-celled state, in which there are no well-defined internal structures, to multicellular forms in which there is a division of labour between different parts of the body. Using nutritional and organizational characteristics Fungi, Plantae and Animalia rank equally in the hierarchy of the five kingdoms (Fig. 1.2).

Fungi are primarily multicellular and multinucleate, that is each cell contains more than one, and frequently several, nuclei and the cells have rigid walls. They

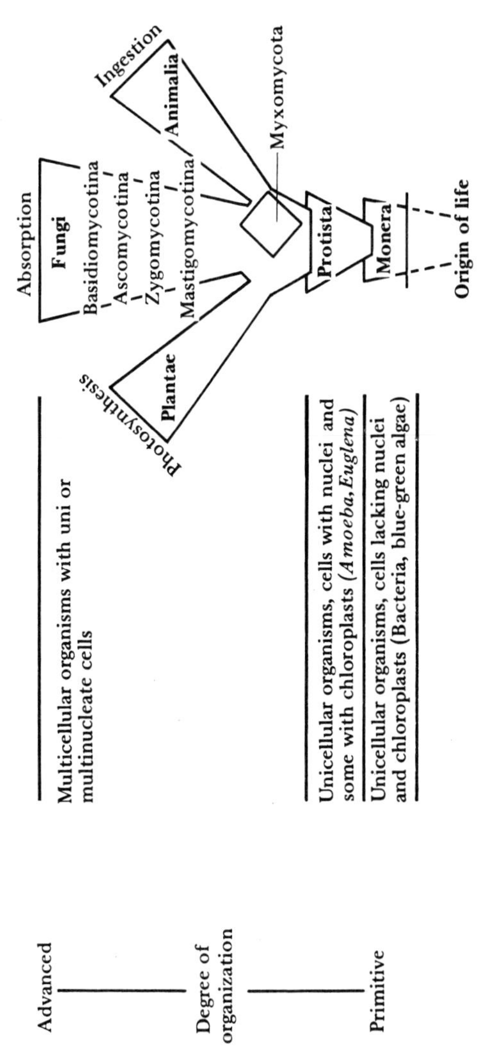

Fig. 1.2 A five-kingdom system for the classification of living things based upon their mode of nutrition and degree of organization. The size of the kingdom Fungi is exaggerated with respect to the other kingdoms. The Myxomycota share fungal and animal characteristics.

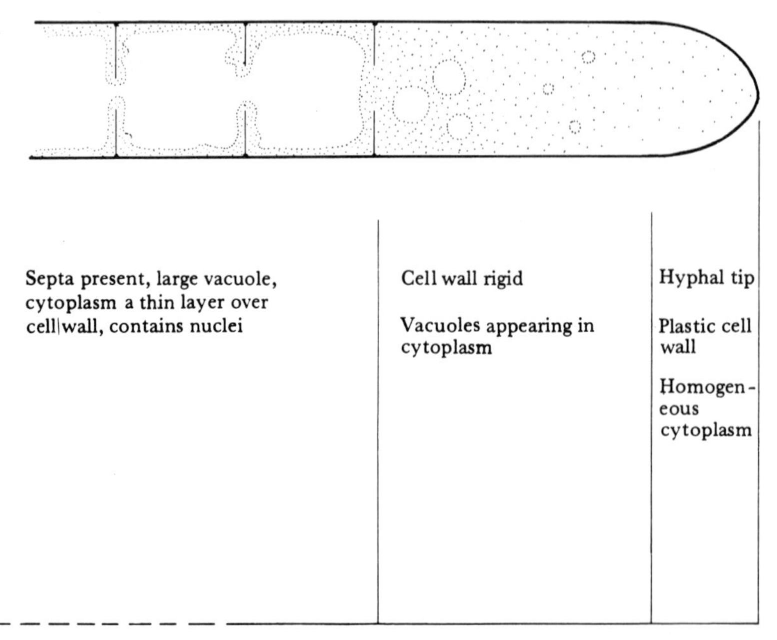

Fig. 1.3 Diagram of the structure of a septate hypha indicating its mode of growth and differentiation.

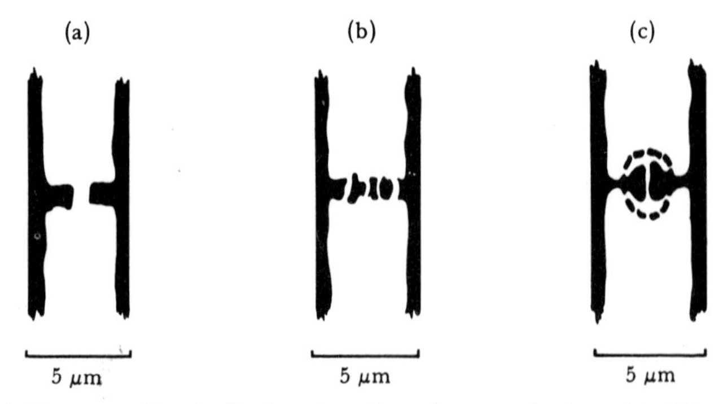

Fig. 1.4 Diagrams of longitudinal sections through septate hyphae with different pore types: (a) simple pore; (b) multiple pores; (c) complex pore typical of Basidiomycetes. Scale, $1\mu m = 0{\cdot}001$ mm.

lack photosynthetic pigments and, since their mode of nutrition is absorption, part or all of their tissues must be immersed in their food supply. In general, differentiation of their vegetative material into distinct, complex organs with separate functions is absent or not well developed, although there is differentiation of reproductive structures and the life cycles of fungi can be complicated. These are some of the broad characteristics of fungi but it is important to remember that kingdoms grade into one another in those areas where degrees of organization within them are at a relatively low level. A number of 'lower' fungi resemble Protista in that they are typically unicellular and uninucleate, while others lack rigid cell walls and behave in many ways like animals. For example, they may be motile and ingest organic matter during certain phases of their life cycle. Some fungi so closely resemble algae, both morphologically and in their pattern of life cycles, that they are considered by certain authorities to be non-pigmented algae. These various areas of overlap are reflections of the possible evolutionary relationships between the various kingdoms and there is no virtue in attempting to draw rigid boundaries through them.

The kingdom Fungi is not a small one either in terms of species or individual organisms. The total number of species is unknown, but those so far described amount to over 50,000 and these are being added to at the rate of 1,000 a year. When it is remembered that most are microscopic, and that the number of mycologists in the world is very small, then these figures must represent only a fraction of those fungi that exist. Estimates of the probable number range from 100,000 to 250,000, the latter figure being comparable with the probable number of existing species of flowering plants. The number of individual fungi is beyond estimation. For instance, in a gram of soil there may be in excess of 100,000 spores.

In size fungi range from the minute to the massive, but the basic structural unit in the vast majority of them is the same. This is the hypha, and each is a microscopic, tubular, branched filament lined with cytoplasm that contains nuclei, and which surrounds a central vacuole filled with a watery sap (Fig. 1.3). Hyphae collectively form the body, or mycelium, of the fungus. Hyphae may lack cross-walls, that is be aseptate, except where reproductive structures are formed; of they may have frequent septa which divide each hypha into multinucleate compartments. When hyphae are septate each cross-wall is usually breached by either a number of pores or by a single, central pore which in some fungi may have a complicated structure (Fig. 1.4). These perforations provide communication between adjacent cells and ensure that all the cells of a septate hypha are connected by means of their cytoplasm. In this the fungi differ from multicellular plants and animals in which the cells are much more independent of one another. Growth of hyphae and hyphal branches takes place only at their ultimate tips which are tapered and have thin, plastic walls. The cell wall immediately behind the tip is continually being added to and made rigid so that the tubular shape of the hypha is not lost (Fig. 1.3). Branching takes place when new growth points arise on the hyphal wall at some distance behind the advancing hyphal tip. Growth of the cytoplasm is maintained by the movement of nutrients into the mycelium through the hyphal surfaces, the mycelium usually being immersed in the substrate upon which it is growing.

Mycelia which arise in nature or are grown in the laboratory usually originate either from a spore or a minute mass of hyphae. From such inocula a small number of apically extending hyphae arise on which lateral hyphae develop so that a young mycelium, or colony, is quickly formed (Fig. 1.5). At the colony margin are the youngest hyphae, lying roughly parallel to one another, which are responsible for the increase in size of the colony. In the central, older regions of the colony, intense lateral branching takes place. If the fungus develops on a thin layer of substrate then the colony grows in only two dimensions and is circular. If it develops while suspended in a liquid medium or grows within a deep, solid substrate then growth takes place in three dimensions and the colony is spherical. Circular or spherical growth forms are almost universal in the fungi, and colonies can theoretically be of infinite size. If the growing margin is not halted by a barrier or by other causes, and if nutrients remain freely available,

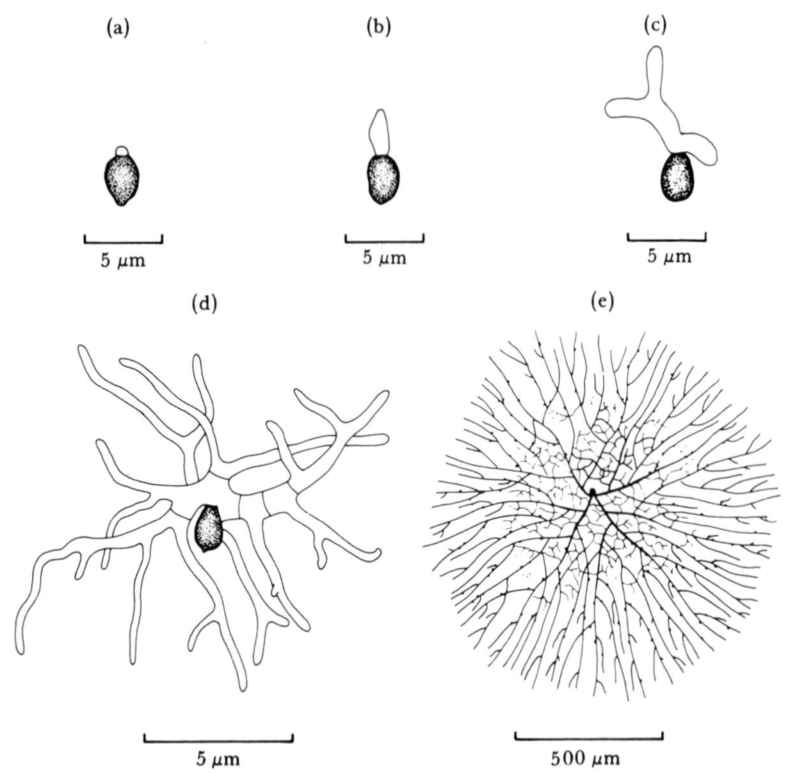

Fig. 1.5 Development of a colony from a single spore: (a) first signs of germination; (b) emergence of germ tube; (c–d) branching of germling hyphae; (e) young circular colony. (After A. H. R. Buller, *Researches on Fungi, Vol. IV*, Longman 1931.)

then there is no limit to the growth of the organism. In practice, of course, colonies remain relatively small since nutrients become limiting, and fungi are also affected by a number of other physical and biological factors that restrict their growth or result in their destruction.

Hyphal branches can fuse with one another and after fusion has taken place the cell walls which separate the cytoplasm of the branches breaks down so that they become interconnected. Branches may either fuse with those from the same parent hypha or with branches from different parent hyphae, so that a colony comprises a network of communicating filaments, and any given part of a colony is ultimately connected to all other parts. A consequence of hyphal fusion is that some fungi are able to build up a range of complex tissues from simple hyphal units. Swelling of the constituent cells of a mycelium, thickening of their walls and rapid growth leading to tight compression of hyphae, can either singly or together lead to the production of tissues with a range of mechanical and biological properties. The most commonly encountered expression of massive tissue formation is to be seen in the production of reproductive structures such as mushrooms and toadstools.

Not all fungi develop a complex mycelium of potentially unlimited size. Indeed, some have such limited development that their bodies, or thalli, consist of a very few cells, or are so reduced in form that the thallus is a single, small cell (Fig. 1.6). The most familiar examples of unicellular fungi are the yeasts

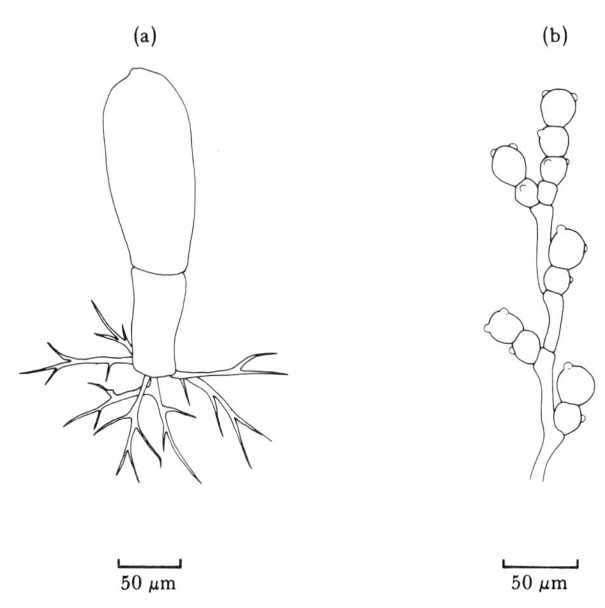

Fig. 1.6 Fungi having thalli of limited size: (a) *Blastocladiella*; (b) *Allomyces*.

whose most remarkable characteristic is an ability for rapid multiplication. A single yeast cell grows until it reaches a certain maximum size at which point one of two things happens depending on the type of yeast involved. In some yeasts—'fission yeasts'—the cell divides so that two daughter cells of more or less equal size are formed (Fig. 1.7). These in turn grow and then divide, this process being repeated as long as sufficient nutrients or space are available to allow it to take place. The population of yeast cells, therefore, doubles regularly with time and since the time taken for growth and division to occur is quite small, sometimes as little as 20 minutes, rapid cell production can take place under suitable conditions. Although each cell is minute, the total mass of material produced can be very large. The cells of other yeasts bud at a point on their wall and the resultant daughter cell grows attached to the mother cell (Fig. 1.7). When the young cell reaches a critical size it separates from the mother cell and may then itself begin to bud. Budding of a cell can take place simultaneously at more than one point on its surface and daughter cells may also begin budding before separation. Small masses or chains of cells are thus produced and, as with fision yeasts, rates of cell production are very high.

Sooner or later during vegetative growth fungi enter a reproductive phase in which dispersive units, usually known as spores, are formed on the mycelium thus enabling the fungus either to spread its progeny to other substrates, or to ensure that it can survive the onset of unfavourable conditions of brief or long duration. In general, fungi can reproduce both asexually, that is without the mediation of a sexual process, and sexually.

Asexual reproduction results in the formation of units which are genetically identical to the parent mycelium. It commonly takes place on hyphae which emerge from the substratum within which the fungus is growing, and it is within or upon these hyphae that asexual spores are produced. Their modes or origin, size, shape, manner of liberation, and dissemination are legion but they are of two basic types, being either naked zoospores that possess flagella and can swim, or non-motile spores which have a rigid cell wall surrounding them. Zoospores

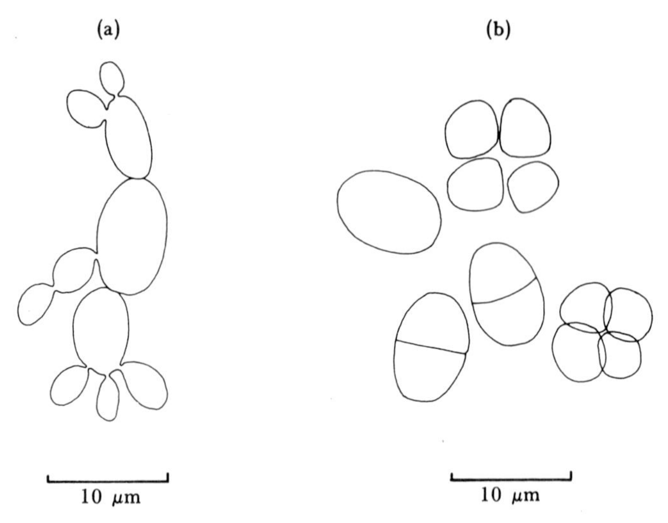

Fig. 1.7 The two modes of yeast growth: (a) budding of cells in *Saccharomyces*; (b) division of cells of *Schizosaccharomyces*.

are produced inside, and are liberated from, specialized hyphae called zoosporangia (Fig. 1.8). In contrast, non-motile spores are formed in a vast variety of ways. They can, like zoospores, be formed within sporangia and they are then known as sporangiospores (Fig. 1.9). Spores that are not formed in sporangia are termed conidia, and the hyphae which bear them are called conidiophores. Conidia can be produced by fragmentation of the conidiophore, by the budding or growth of cells which form the apex of the conidiophore, or arise from special growth points located on the conidiophore. Depending on their origin and subsequent development pattern, conidia may be produced singly or in chains, may be small or massive, unicellular or multicellular, and may contain one or many nuclei (Fig. 1.10).

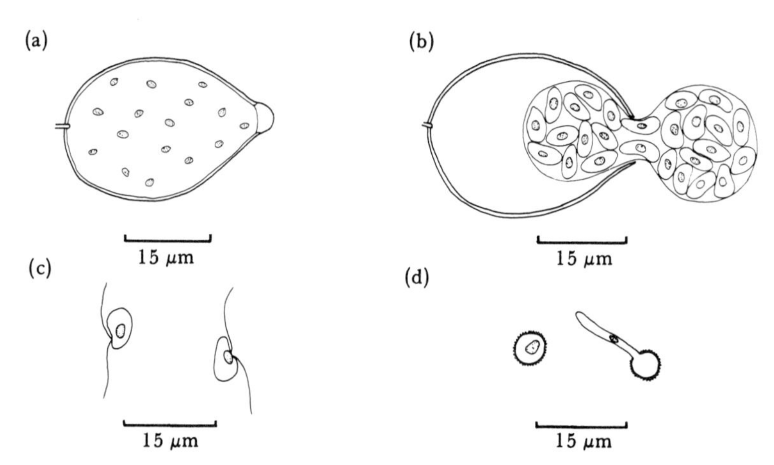

Fig. 1.8 A zoosporangium and zoospores of the phycomycetous fungus *Phytophthora*: (a) a multinucleate zoosporangium; (b) emergence of uninucleate zoospores; (c) biflagellate zoospores; (d) an encysted zoospore and a cyst producing a germ tube.

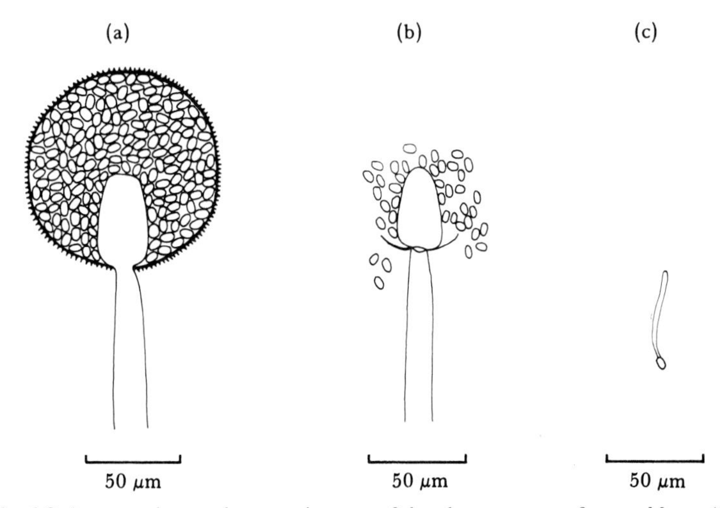

Fig. 1.9 A sporangium and sporangiospores of the phycomycetous fungus *Mucor*: (a) a sporangium containing non-motile sporangiospores; (b) fragmentation of sporangium wall to release the spores; (c) spore producing a germ tube.

Zoospores, non-motile sporangiospores and conidia resemble one another fundamentally in that they all contain a fragment of the parent's cytoplasm and have nuclei which carry genetic information identical to that possessed by the parent. When any asexual spore subsequently gives rise to a new mycelium this will therefore be genetically identical to the parent mycelium. The majority serve the same broad purpose, which is to spread the fungus rapidly and efficiently to new locations. Some asexual spores have a relatively short life, sometimes as brief as a matter of hours, and rapidly lose their viability. This is because they have only a small food reserve within them and are also susceptible to the adverse effects of desiccation, particularly if their walls are thin. Statistically, the chances of a particular spore reaching a favourable habitat in a viable condition are very small. However, asexual spores are usually produced in enormous numbers, so that for any particular species some spores will always arrive somewhere where they can establish a new mycelium, despite the obviously high wastage rate. When a suitable substrate is contacted spores germinate. If the spore is a zoospore its flagella are first withdrawn into the cytoplasm, or are shed, and then the naked cell rounds off and secretes a wall around itself (Fig. 1.8). Zoospores, non-

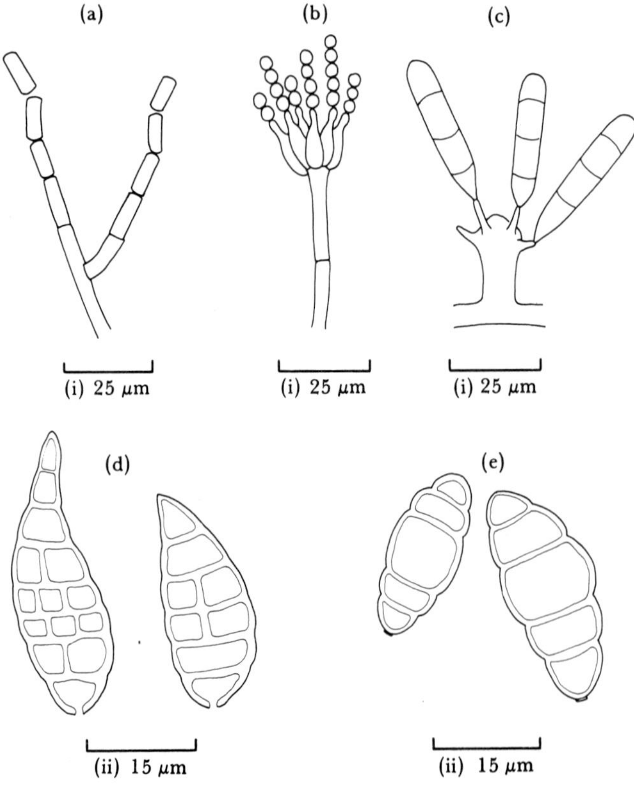

Fig. 1.10 Conidia and conidiospores of some Fungi Imperfecti: (a) *Geotrichum*; (b) *Penicillium*; (c) *Dactylaria*; (d) *Alternaria*; (e) *Curvularia*. (a)–(c) to scale (i), (d)–(e) to scale (ii).

motile sporangiospores and conidia generally all germinate in the same way. A small hypha, the germ tube, arises from the spore and grows through the substratum (Figs. 1.8 and 1.9). This branches, the branches grow, each branching in turn, so that eventually either a typical colony or a limited thallus is produced, depending on the species. More than one germ tube may emerge from a unicellular spore, and if a spore is multicellular a germ tube can usually arise from each of its constituent cells.

In contrast, *sexual reproduction* involves the fusion of two nuclei which are genetically different from one another and results in the production of daughter nuclei within which the genetic characters of the two original parent nuclei have become reassorted and combined. A mycelium derived from one of these daughter nuclei will thus have characteristics which differ from those possessed by the parent nuclei. The ways in which sexual nuclei are brought together and the subsequent events that follow fusion are extremely diverse. They may both be carried in motile cells, or only one cell may be motile, or both may be non-motile and remain attached to the mycelium (Fig. 1.11). Fusion may take place between nuclei from separate mycelia or between those from the same parent mycelium. Where two mycelia are involved these may be either distinguishable sexually as male and female or resemble one another so closely as to appear identical (Fig. 1.11). Some sexual spores are frequently thick-walled and have abundant food reserves so that they are durable and can survive adverse conditions for considerable periods, often for several years.

Some fungi can produce resistant structures other than sexual spores, for example thick-walled, multinucleate chlamydospores which are formed simply by the rounding off of one or more hyphal segments. Other species commonly and characteristically produce sclerotia. These are formed from small masses of hyphae, the cells of which swell and fuse together to form a large regular or irregular multicellular body. The outer layers of this may become compacted into a hard, dark rind enclosing softer inner tissues, the cells of which are packed with food reserves.

Although the fungal kingdom contains organisms which are very diverse in size, organization, and complexity of reproductive behaviour, it is a relatively easy matter to separate them into two major divisions, using as a criterion the nature of their vegetative structure (Table 1.1). Division I, the Myxomycota, comprises organisms that are animal-like in many ways and some opinions even favour excluding them altogether from the fungi. Division II, the Eumycota, contains all those species that are generally regarded as being typical fungi, and these can be distributed among five subdivisions, depending on their vegetative and reproductive characteristics. These large, primary divisions and subdivisions form the basis for a more detailed classification of fungi, but there is by no means full agreement as to how the kingdom should be divided, and no single scheme is universally accepted. In nearly all books concerned with fungi, and among most mycologists, the Mastigomycotina and Zygomycotina are collectively referred to as the Phycomycetes, the Ascomycotina as the Ascomycetes, the Basidiomycotina as the Basidiomycetes, and the Deuteromycotina as the Fungi Imperfecti. These are old names of long-standing, which means that, justifiably, there is great reluctance to abandon them.

Each subdivision contains a number of classes, each of which comprises a number of orders. An order can usually be divided into families, each family into genera, and each genus into species. As each unit within the classification becomes smaller, more and more specialized groups of characters are used to

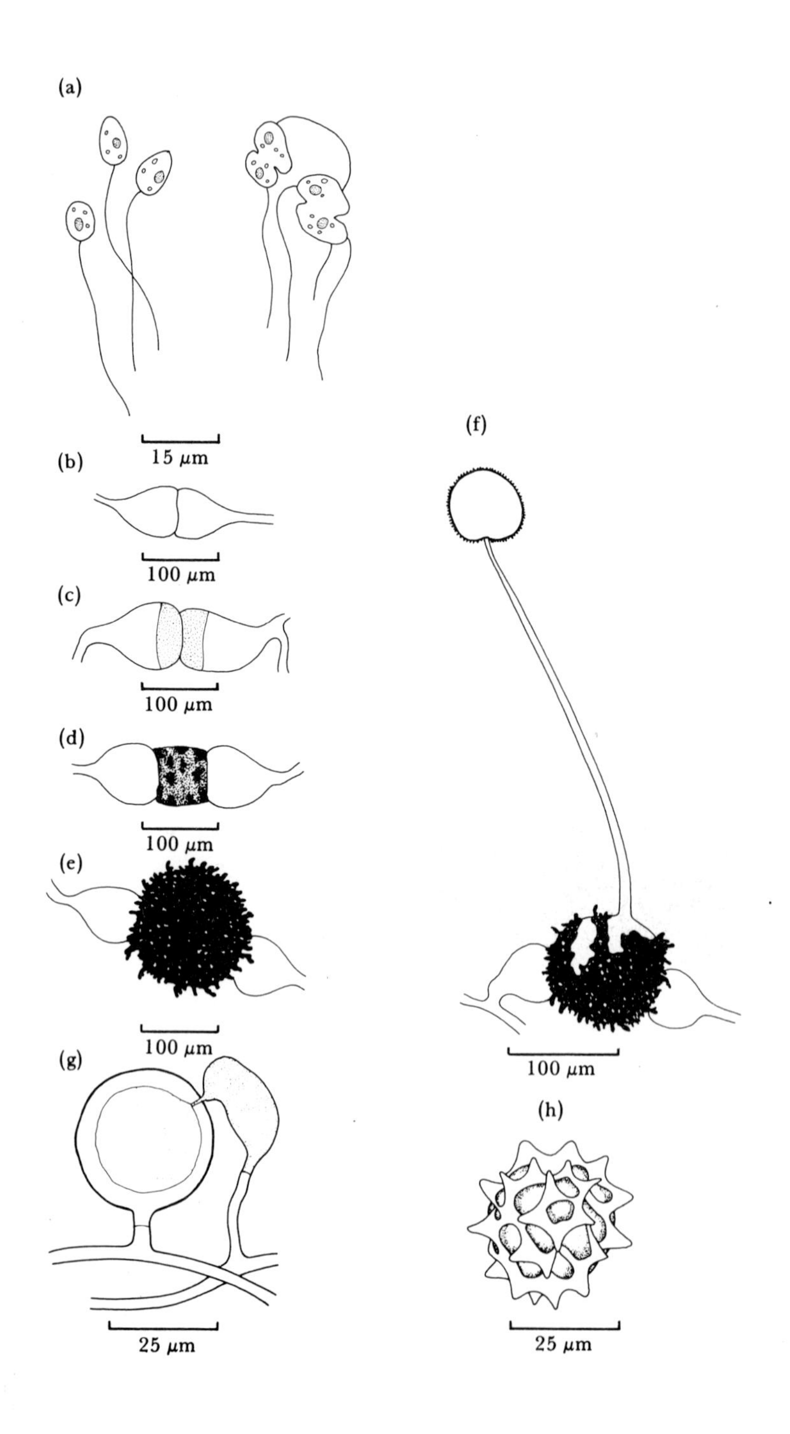
(a)
15 μm
(b)
100 μm
(c)
100 μm
(d)
100 μm
(e)
100 μm
(f)
100 μm
(g)
25 μm
(h)
25 μm

Table 1.1 Divisions and subdivisions of the fungi with some commonly encountered representatives of them.*

		Divisions	
		I Myxomycota	Consist of naked, amoeba-like cells, aggregations of these cells, or large, multinucleate masses of cytoplasm. Motile at some stage, nutrition sometimes by ingestion. Slime fungi.
		II Eumycota	Typically mycelial, but may have a thallus of a few cells or be unicellular.
		Subdivisions	
(Phycomycetes)	=	Mastigomycotina	Unicellular or mycelial, hyphae aseptate, spores and gametes commonly motile. Fungi growing on fish in aquaria (*Saprolegnia*); *Phytophthora infestans* causing potato blight.
		Zygomycotina	Mycelial, hyphae aseptate, spores and gametes non-motile. Pin moulds on food and leather (*Mucor*, *Rhizopus*)
(Ascomycetes)	=	Ascomycotina	Mycelial, hyphae septate (but yeasts unicellular), sexual spores (ascospores) produced within special cells (asci). Brewer's and baker's yeast (*Saccharomyces*); cup fungi on wood and burnt ground; apple and pear scab (*Venturia*).
(Basidiomycetes)	=	Basidiomycotina	Mycelial, hyphae septate (but yeasts unicellular), sexual spores (basidiospores) produced on special cells (basidia). Mushrooms, toadstools, puffballs; dry rot fungi; bracket fungi.
(Fungi Imperfecti)	=	Deuteromycotina	Mycelial, hyphae septate (but yeasts unicellular), sexual reproduction absent. Green and blue moulds on food and leather; the blue and white moulds of fungally-modified cheeses (*Penicillium*, *Aspergillus*); 'Ringworm' fungi of man and animals (*Trichophyton*, *Microsporum*)

*Subdivision names in parenthesis are frequently-used alternatives.

Fig. 1.11 Some modes of sexual reproduction in phycomycetous fungi: (a) fusion of uniflagellate gametes; (b)–(e) fusion of non-motile gametangia of equal size in *Mucor* and formation of a thick-walled zygospore; (f) germination of a zygospore to give rise to a sporangium; (g) fusion of a large (female) and a smaller (male) gametangium in *Pythium*; (h) a thick-walled oospore resulting from this fusion.

delineate or define it. Again, opinions frequently differ as to the limits and characteristics of each group from the class to the species. It should be noted that yeasts appear in the Ascomycetes, Basidiomycetes, and Fungi Imperfecti. This is because the term 'yeast' refers to a special mode of growth and does not describe a particular, special assemblage of fungi. A yeast is a fungus which, irrespective of its mode of sexual reproduction, is unicellular and grows through the process of fission or budding. However, many species that are normally mycelial adopt a yeast form if they are growing under conditions of environmental stress—for example, if the substrate is not suitable for mycelial growth or if the temperature is too high. The Fungi Imperfecti contain a large variety of species whose common attribute is the lack of a sexual phase in their life cycle. Such a phase may be absolutely absent, because the fungi are genetically incapable of this kind of reproduction, or it may only apparently be absent, either owing to a lack of critical observation or because the fungi have not been grown or observed under conditions favourable for sexual reproduction. This subdivision is therefore to some extent a dumping ground for species that cannot legitimately be placed in one of the others. When a sexual stage is discovered, as they often are, a species can then be moved to its correct place in the appropriate subdivision.

The classification of fungi might seem to be a purely intellectual exercise but a good system of classification, or a number of schemes of equal merit, is desirable for two basic reasons. First, it is essential that a degree of order be established that will allow all workers in the same field to know exactly which organism is being referred to by any other worker. Secondly, the degree of order that is maintained should ideally mirror the probable evolutionary relationships between the various groups of fungi, large or small, that exist within it. For example, there should be indications as to which are primitive and which are advanced groups, or which groups have a low degree of organization and which are complex. It is obvious that, although many admirable systems of classification are available, no single scheme is ever likely to find universal acceptance. Order is imposed by the classifier and nature is not always necessarily neatly ordered. This is especially true when attempts are made to draw clear-cut lines through areas in which groups of fungi grade into one another. In addition, although evolution moves from the simple to the complex it can also occur in the reverse direction, so that simplicity of form does not always represent primitiveness. Finally, all criteria used in classification are essentially subjective and any classification of fungi is valuable only so long as its unavoidable fundamental rigidity is not imposed too strictly.

It is impossible to be unaware of the complexity of plants and animals simply because the greater number of them are macroscopic and their structure and behaviour patterns are easily observed. Because of their small size this situation is not the case with the vast majority of fungi. Even so, although in comparison with plants and animals fungi are physically relatively simple, their behaviour can be more complicated than might be expected. In addition, the manner in which they interact with other microorganisms or with plants and animals can have far-reaching results, frequently on a scale which directly or indirectly affects man's way of life. It is to a consideration of some of the effects of fungi on man's environment, both physical and social, that the remainder of this book is addressed.

2

Nutrients and niches

.... We
Diet on water,
On crumbs of shadow,
Bland-mannered, asking
Little or nothing.
Sylvia Plath, *Mushrooms* 1967

Within the fungi are to be found a great variety of behaviour patterns. There exists a whole gamut of relationships with other organisms and many species possess an ability to occupy peculiar ecological niches. This behavioural diversity is matched by the nutritional versatility of fungi, so that as a group they can use a very wide range of substrates from which to draw their sustenance. In any particular situation the way of life of a fungus is to a large extent determined by the kinds of nutrients that are available to it. It is, therefore, obvious that an understanding of the nutrient requirements of a fungus, and of the manner in which it satisfies these requirements, is fundamental to an appreciation of why it grows where it does and why it behaves as it does.

Fungi have the same basic nutrient needs as all other living things but differ in the ways in which they satisfy them. They have a major requirement for organic carbon compounds, usually in the form of carbohydrates, the carbon from which is used to manufacture cytoplasm or structural compounds for cell growth. The hydrogen from the carbohydrates is used to provide the chemical reducing power necessary for the multitude of biochemical reactions that take place within the cytoplasm. Carbohydrates are broken down into their component parts during the process of respiration and energy is released that can then be utilized by the fungus. In addition to carbon compounds, fungi require a source of nitrogen, which is an essential component of proteins, and also mineral elements such as phosphorus, potassium, iron, magnesium, zinc and sodium, usually in the form of chloride, sulphate and phosphate salts. Vitamins and other growth factors are also necessary for the normal functioning of fungal cells, some fungi being able to synthesize their own vitamins, others being dependent on an external source.

Photosynthetic organisms are autotrophic, i.e. they can manufacture their own carbohydrates and other organic compounds from atmospheric carbon dioxide. Unlike these organisms, and in common with other non-photosynthetic life forms, fungi are heterotrophic and do not have this ability. The obvious consequence of heterotrophy is that fungi must obtain their carbohydrates from a source outside their own cells and since, with few exceptions, they cannot ingest solid organic particles, carbohydrates and other nutrients must be in solution so that they can be absorbed. There are three broad modes of life which fungi have adopted in order to satisfy their carbohydrate needs. First, some are parasitic on plants or animals and obtain the necessary carbon compounds by destructive exploitation of the host organism. Secondly, others have become symbiotic with plants or animals, i.e. they live with them in prolonged or permanent intimate contact, and obtain carbon from their hosts without doing the latter damage of any serious kind. Finally, most fungi are saprotrophic and live on dead, decaying organic matter derived from plant or animal tissues.

The relationship between symbiotic and parasitic fungi and their hosts have for a long time provided an area of research activity. More recently, the possible mechanisms by which nutrient exchange is effected, and the consequences of such an exchange for both fungus and host, have been subjected to intense scrutiny. In terms of effort, the activities of saprotrophic fungi have probably received no less attention, but their fundamentally important role in nature and their present and future importance to man seem, in the main, to excite much less general interest. Aspects of symbiosis and parasitism are dealt with in the appropriate place in subsequent chapters and it is the intention here to consider some of the major activities of saprotrophs. The biosphere, in which all organisms on earth exist, is maintained in a state of equilibrium because the material of living things is constantly being cycled. Saprotrophic fungi play a ubiquitous and active part in this process since they can efficiently and rapidly transform and utilize the dead remains of other organisms. In a wide ecological context this biodegradation is of obvious benefit to all organisms including man. Unfortunately, saprotrophic fungi also indiscriminately transform natural and manufactured organic materials which are in the service of man, so that the immediate result of such biodeterioration is not beneficial.

A major process in which saprotrophs play a fundamental role is the carbon cycle, the carbon locked up in dead organic matter being utilized by them, and ultimately released from them in the form of respiratory carbon dioxide. This is then used by autotrophs to manufacture once more those organic compounds upon which all life depends. Because the amount of biologically utilizable carbon on the earth and in its atmosphere is not infinite, this constant cycling of carbon is essential for the maintenance of life. Nitrogen and minerals are cycled in a similar way.

The recycling of organic matter through natural processes takes place on a vast scale in both the terrestrial and marine environment, although fungi are probably not important decomposer organisms in the open sea. On land the most obvious example of recycling is the regular fall of leaf litter, approximately half the weight of which consists of carbon. In mature forests a state of equilibrium has been reached between the annual litter fall and the amount of material that has accumulated on the forest floor (Table 2.1). By examining the amount of these two kinds of litter in any situation an idea of how rapidly decomposition must take place to maintain this equilibrium can be obtained. Decomposition rates are rapid in tropical rain forests where high temperatures and moisture levels favour intense activity of decomposer organisms, among which the saprotrophic fungi are an important group. In drier, more temperate, regions decomposition is slower, particularly in pine forests, where the needles do not provide good material for the decomposers.

Fresh, dead organic substrates are composed of both soluble and insoluble compounds. The soluble compounds are available relatively readily and are quickly absorbed by the fungi, leaving the insoluble components which must then be degraded. Degradation is brought about by the action of enzymes that are produced within the hyphae and diffuse out of them into the substrate. These break down the insoluble compounds into soluble breakdown products which then diffuse into the hyphae through the cell walls. Since hyphae are characteristically immersed in the substrate, they are consequently bathed in the soluble nutrients released by the action of their extracellular enzymes. The effect of the latter also allows fungi to move through solid organic matter by eroding channels that act as routes for hyphal growth, or by rendering it soft enough to be penetrated by means of the mechanical pressure exerted by extending hyphal tips.

Table 2.1 Annual litter fall and constant amount of accumulated litter in various kinds of forest of temperate and tropical zones.

	Annual rainfall (cm)	**Average temperature** (°C)	**Annual litter fall (kg per hectare)**	**Accumulated litter (kg per hectare)**
Tropical forests				
Mixed rain forests	170	24·5	12,300	4,300
Broad-leaved rain forests	825	25·5	7,560	4,500
Temperate forests				
Mixed woodlands	85	10	3,100	3,600
Pine woodlands	75	10	2,800	167,700

It is impossible to enumerate here all the natural substrates that fungi can degrade and grow upon, but four of the most common insoluble components of such substrates are cellulose, lignin, chitin and keratin. Cellulose and lignin are characteristic of plant debris; chitin forms the major part of insect exoskeletons; while hair and feathers are composed of keratin. The basic unit of each compound is a soluble molecule and these molecules are linked together to form simple or complex chains (Figs. 2.1 and 2.2). If the chains are short, i.e. composed of only a few molecules, then they are soluble, but as their length increases their solubility decreases. Extracellular enzymes break down the atomic linkages between molecules, and long chains are fragmented in this way so that their constituent molecules are made available for absorption by the fungus. The ability of an enzyme to break a linkage is related to the physico-chemical nature of that linkage, so that whether or not a particular enzyme will degrade a particular compound depends on the manner in which that compound's component molecules are joined together. This means that enzymes are normally to some extent specific. For example, cellulolytic enzymes which can cleave the linkage between glucose molecules in cellulose cannot break the unions between propanoid units in lignin. Whether a particular fungus can attack a particular natural substrate is thus obviously determined by the chemical composition of that substrate in relation to the enzyme-producing properties of the fungus, the success of the fungus in colonizing the substrate depending on its capability to produce the necessary specific enzymes.

By no means all fungi are able to make use of all components of natural organic substrates, so that progressive decomposition of these substrates is usually accompanied by a characteristic pattern of colonization by different groups of fungi.

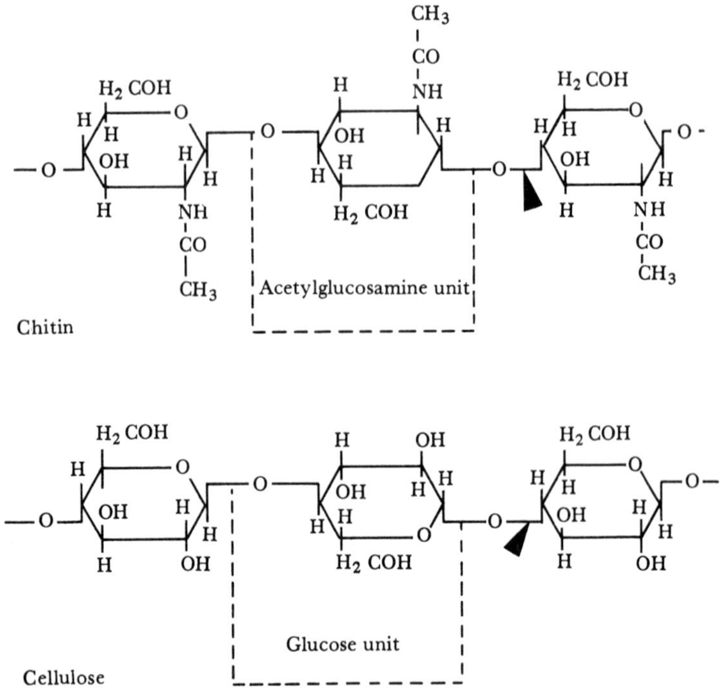

Acetylglucosamine unit

Chitin

Glucose unit

Cellulose

Fig. 2.1 Structure of chitin and cellulose showing the manner of linkage of their basic units. Arrows indicate the points at which enzymes act to break the chains into smaller units.

Freshly-fallen leaves contain quite large amounts of soluble carbohydrates, cellulose, and to a lesser extent lignin. Those fungi that lack cellulases and lignin-degrading enzymes will only be able to utilize the soluble components of such leaf tissues. These fungi colonize the leaves first, rapidly use up the carbohydrates, and then die or produce spores. Meanwhile species that can also utilize cellulose and lignin have invaded the leaves. Cellulose is rapidly broken down but lignin is more resistant so that finally only fungi with the ability to degrade lignin remain active in the decomposing leaves. Many special ecological groups of fungi are able to occupy peculiar organic substrates that are denied to other fungi which lack the necessary enzymatic equipment to exploit them. For instance, relatively few species are able to use keratin as a carbon source, but these species are abundant in situations where keratin substrates are common. They form a characteristic fungal flora in birds' nests, the pellets of birds of prey and the quills of living and dead hedgehogs.

The natural decomposing activities of fungi are ecologically essential on a global scale and, in addition, in restricted and carefully controlled situations, these activities can be made use of commercially for a variety of purposes. Mushroom growing, the production of fungally-modified cheeses, the manufacture of fermented drinks and many more processes involve the use of saprotrophic fungi for the degradation or modification of selected substrates. They are also of obvious use in the breakdown of industrial and domestic wastes and help to

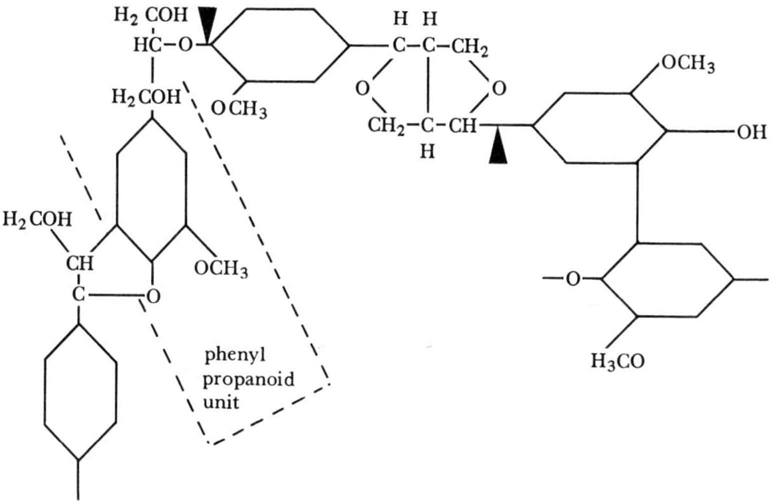

Fig. 2.2 The possible structure of lignin, a complex network of linked phenylpropanoid units. Arrows indicate the points at which lignin-degrading enzymes might possibly act.

reduce the disposal problems normally associated with these materials. On average, 55 percent of domestic refuse in the United Kingdom consists of biodegradable components with a high cellulose content, mainly in the form of paper and vegetable waste. A large proportion of it is therefore eminently suited to microbial recycling, the end product being useful compost.

An outstanding and characteristic feature of the composting of organic waste is the very rapid rise in temperature which takes place at the centre of the material. The maximum temperature reached and the duration of high temperature conditions vary with the nature of the substrates being used, but a peak of 65–70°C is not uncommon, and after this a temperature of 50°C may be maintained for several weeks. It has long been known that heat is generated within composting materials as a result of the activity of microorganisms, among which fungi are an important group. Most fungi, particularly when in the form of spores, can survive extremes of heat and cold, but for the most part they are mesophilic, i.e. they grow optimally at 20–30°C and usually cease to grow below 5°C and above 40°C. Those species which inhabit the inner regions of compost heaps are thermophilic, having a maximum temperature for growth of above

50°C and a minimum temperature at or above 20°C. Although relatively few fungi are thermophiles they are efficient decomposers at these high temperatures, having a particularly well-developed ability to degrade cellulose. Thermophiles are, of course, not the only fungi present within compost and, away from the zone of most intense heat, a wide range of mesophiles will also be active.

Not unnaturally, the propensity of thermophilic fungi to grow at temperatures that are lethal to most other organisms has aroused interest. Death at elevated temperatures is due to a number of factors, but perhaps the most important is the effect of heat on proteins. These are normally thermolabile and lose their stability with increasing temperature so that cytoplasm is destroyed and enzymes, which are also proteinaceous, are inactivated. In addition, essential fats contained within cells melt and the minute organelles that control cell function become disorganized. Thermophiles must in some way be able to prevent one or all these events from occurring or be able to alleviate the effects of them on cell metabolism. There are several means by which they might achieve this, although it has not been shown experimentally which mechanisms operate in which species. First, the structure of their proteins could be such that these have a high degree of thermostability and can thus retain their function even at high temperatures. In the absence of such thermostability it is possible that the fungi can manufacture proteins at a rate equal to the rate at which they are being destroyed. Secondly, their fats may contain certain fatty acids which raise their melting point sufficiently for the maintenance of their integrity. Finally, the ultrastructural elements or organelles of thermophilic fungi may, like their proteins, be thermostable.

While it is relatively easy to envisage a number of mechanisms that might enable thermophiles to grow at high temperatures, it is more difficult to account for their high minimum temperature for growth. Why they should be prevented from growing at ordinary temperatures at which mesophiles thrive is not known. It has, however, been suggested that one or more heat-stable enzymes are inhibited by a somewhat less heat-stable factor. Enzyme action cannot then commence until the temperature rises high enough to destroy the inhibitor, so that metabolism is only possible at temperatures above which inhibition is removed.

Lest it be thought that fungi are either mesophilic or thermophilic, it should be pointed out that fungi also live in very cold environments, and in nature they have consistently been found in both Arctic and Antarctic habitats. Some of these species may be mesophiles that are merely tolerating the extreme cold in an inactive condition and which can grow only during the relatively warm periods of the summer months, but others are psychrophiles and can grow at temperatures below zero. One psychrophilic yeast has been recorded as being capable of growth at −34°C. Some cold-tolerant mesophilic species are found in cold, man-made habitats and many cause the spoilage of foodstuffs that are held in refrigerated storage. *Cladosporium herbarum* is the best-known example and this species will grow on meat stored at −6°C. Being mesophilic it grows slowly at sub-zero temperatures but a rise in temperature to 2°C favours profuse growth. This is below the normal operating temperature of many domestic refrigerators.

Cladosporium herbarum is but one example of a multitude of fungi that cause the biodeterioration of essential or economically important resources. The large-scale loss of food commodities and the destruction of structural and other materials through biodeterioration is in many cases due to unrestrained saprotrophic growth. All property owners are aware of the dangers of dry rot and wet rot in construction timber. These are caused by *Serpula lacrymans* and *Con-*

iophora cerebella, respectively, which are basidiomycetous species with a well-developed ability to break down the components of wood and so reduce its mechanical strength. The economic importance of these two fungi may be judged from the surprising frequency with which they occur in building timbers. It has been estimated that in the United Kingdom alone 5–28 percent of all dwellings contain some degree of dry rot, and that wet rot is present in 17–30 percent. Replacement of diseased timbers must account for a significant proportion of the United Kingdom's total annual import expenditure on timber, which is currently in excess of £200 million.

Some fungi that cause serious forms of biodeterioration occupy strange ecological niches. In tropical climates, especially those where high humidity is a normal condition, the lenses of optical instruments and the glass parts of scientific equipment are often rapidly damaged by the growth of fungi upon them. Conditions of high humidity result in the condensation of films or droplets of water on the glass in which fungi can grow. As a source of energy they can utilize either the cementing materials used in the construction of the glass components, or carbon compounds contained in atmospheric dust particles which become suspended in the water. Growth may only be sparse, but hyphae grow in close contact with the glass and frequently cause such severe etching of the surface that its optical properties are badly affected.

Recently, considerable interest has been evoked by reports that fungal mycelium can be found growing in storage tanks and in aircraft fuel tanks containing kerosene-based fuels. This presents an obvious potential danger in that pipes and valves might become plugged with hyphae, perhaps even during the flight of an aircraft. It has, in addition, been found that where such tanks are constructed of unlined aluminium alloy the presence of fungal growth is associated with corrosion of the tank walls. The main organism involved is an ascomycetous species, *Amorphotheca resinae*, the 'kerosene fungus' or 'creosote fungus'.

Amorphotheca resinae is a common soil fungus and is widespread in nature. It also frequently occurs in resinous wood and in timber that has been treated with preservatives containing creosote. A characteristic of this species is that it is unable to compete for simple, soluble carbohydrates with other microorganisms. In the soil it seems to occupy an ecological niche in which it can utilize carbon compounds that are not available to the great majority of bacteria and fungi and so avoids competition. These carbon compounds appear to be hydrocarbons. Its occurrence in resinous wood and creosote-treated substrates could obviously be due to its being able to utilize resins or creosote as carbon sources. Alternatively, these compounds might inhibit the growth of other organisms while not affecting *A. resinae* so that competition is removed and this species can then use the simple carbohydrates which it is normally denied. What cannot be refuted is that *A. resinae* is able to utilize kerosene-based fuels, which are composed of hydrocarbons, as carbon sources.

When its spores are sown in aviation fuel they fail to germinate in the absence of water, but they do remain viable. If water is subsequently added to the fuel the spores germinate, and within two weeks a mycelium is formed at the fuel–water interface. Amazingly, provided water is available, *A. resinae* can also use kerosene vapour as an energy source, and the presence of liquid fuel is not essential for hyphal growth. The spores of the fungus have been detected in the atmosphere and it is easy to visualize how they might find their way into fuel tanks. If these contain even slight traces of water, due to condensation or other causes, then growth of the fungus within them at fuel–water interfaces is possible. Where mycelium is in contact with aluminium alloy the metal is corroded within the area of contact and the degree of this corrosion can be severe. It has been shown that pieces of aluminium foil placed in contact with *A. resinae* can lose up to 27 percent of their weight within 6 weeks. The way in which the fungus brings about corrosion is interesting. During growth it produces organic acids that diffuse into the water drop or film in which it is developing. If the water is in contact with aluminium the dissolved acids act as an electrolyte for the establishment of an electric cell, the fungus acting as the anode and the aluminium acting as the cathode, and quite high voltage differences can be detected between the fungus and the metal. Aluminium therefore passes into solution in the electrolyte and the tank wall is in this way locally eroded.

While the occupation of strange habitats by a number of fungi or groups of fungi can be explained on the basis of their mode of nutrition, or their inability to compete with other microorganisms, this is not always the case. It has already been mentioned that many fungi have become adapted during evolutionary time so that they can occupy and thrive in ecological niches which are characterized by extreme environmental conditions. In these instances the ability to succeed in such situations is determined by physiological properties of the fungi other than those connected with nutritional processes. Not only are the environmental stresses acting within these niches inimical to the growth of the majority of fungi but they also preclude the growth of the other organisms.

For example, fungi obviously have an absolute requirement for a source of water, but many can grow in conditions of low water availability. Water may be unavailable to a fungus for two reasons. First, it may be entirely lacking, that is physically unavailable. Secondly, it may be abundant but contain such high levels of dissolved substances that the osmotic mechanisms which govern water uptake into living cells cannot operate. It is thus physiologically unavailable rather than physically unavailable. An example of this situation is found in preserved foods such as jam and bottled fruit where mould growth within them is prevented by their high sugar content. The growth of many species is not prevented by the unavailability of water, since they can absorb water vapour from the atmosphere as long as the relative humidity is above 70 percent. Some are also osmophilic, i.e. they are adapted to, and grow much better on, substrates which have high sugar contents.

Occasionally, species that are capable of growth in dry situations can be serious food spoilage organisms, particularly on cereal grains. Harvested grains are inevitably contaminated with fungal spores. They may also contain, beneath their outer layers, hyphae of fungi that have grown within them in a limited way during seed development. These contaminating fungi require free water or a high relative humidity for active growth, so that if the grain is carefully dried before storage they cannot develop within it. For this reason its moisture content must at least be reduced to below 20 percent of its total weight, and in practice the moisture level in bulk-stored grain is below 13 percent. However, some of the contaminating fungi may be species that can grow in low moisture conditions, and should the moisture level rise to a fraction above 13 percent then they will become active. Although their development is not rapid they produce water from their respiratory processes so that their growth accelerates and other fungi, less tolerant of low moisture conditions, are permitted to grow. Localized increases in moisture content of the grain together with increasing metabolic activity of the fungi cause heat to develop. This spoils the grain directly or by allowing the development of thermophilic species. This self-catalytic process of moisture increase and heating may lead to the complete loss of large quantities of stored cereals.

The activities of fungi within organic materials can be either useful or detrimental, depending to a large extent on whether their substrates are considered to be waste or valuable resources. The exploitation of organic material in the environment by fungi is not, quite obviously, restricted to dead organic matter or to stored organic products, and some of their most dramatic effects are to be found in situations where they are interacting, either beneficially or harmfully, with other living organisms. An obvious example is the destruction of plants, particularly man's crop plants, by parasitic fungi.

3

Crop diseases and natural catastrophes

On July 27 I passed from Cork to Dublin, and this doomed plant bloomed in all the luxuriance of an abundant harvest. Returning on August 3rd I beheld with sorrow one wide waste of putrefying vegetation.

Eye-witness of the Irish potato famine, 1846

Plant diseases caused by fungi are multitudinous, new diseases frequently arise and fresh manifestations of well-known diseases are continually emerging. All plants, from minute algae to giant forest trees, have fungi that are able to attack them and commonly destroy them. A major restraint on the productivity of both wild and crop plants is the vast array of rusts, blights, mildews, wilts, and rots to which they are all to some degree susceptible.

Diseases of crops on the conspicuous scale now familiar to us must have made their appearance soon after cultures based on relatively large, permanent agricultural units became established, and numerous references to fungal diseases occur in ancient writings. 'The LORD shall smite thee with a consumption ... and with mildew' it states in Deuteronomy, and it is not difficult to conceive of the unimaginable fears experienced by such peoples on seeing a vital crop being destroyed by some agent beyond their understanding. Nor is there much room for complacency today since, despite our detailed knowledge of the causes and effects of fungal diseases, we are separated from such agricultural disasters by the thinnest of barriers which is only maintained by constant vigilance, research and prophylactic measures. Inevitably this barrier is occasionally breached, sometimes with very serious consequences.

Plant diseases are potentially most dangerous when they strike at staple carbohydrate- or protein-rich crops, these being principally cereals, rice and the potato. If plant diseases were allowed to run their course unchecked the result would be catastrophe. As a warning there is the Irish potato famine of 1845–9 which was caused by the phycomycetous species *Phytophthora infestans*. This fungus destroys the foliage of the potato plant, thus bringing about cessation of tuber growth. The disease is termed 'late blight' and the fungus can also infect the tubers, causing them to rot during storage, so that under some conditions complete loss of a potato crop either in the field or subsequent to harvest can result from the activities of the fungus. Late blight first appeared in Europe in 1844, probably after being imported with new varieties of potatoes either from Central or South

America (Fig. 3.1). It reached Ireland in 1845 and during 1846–7 it did immense damage. Its high rate of spread (about 80 km a week), the rapidity of its effect on the crop and the lack of knowledge concerning cause or prevention of the disease had inevitable and well-documented results. In 1841 the population of Ireland was over 8 million, three-quarters of whom relied on the potato to provide more than half of their annual food intake. By 1851 approximately one million people had died of direct starvation, or of disease resulting from starvation, and a further million had emigrated. About 250,000 of the latter died within a year of leaving Ireland, mainly as a result of ill-health caused by malnutrition.

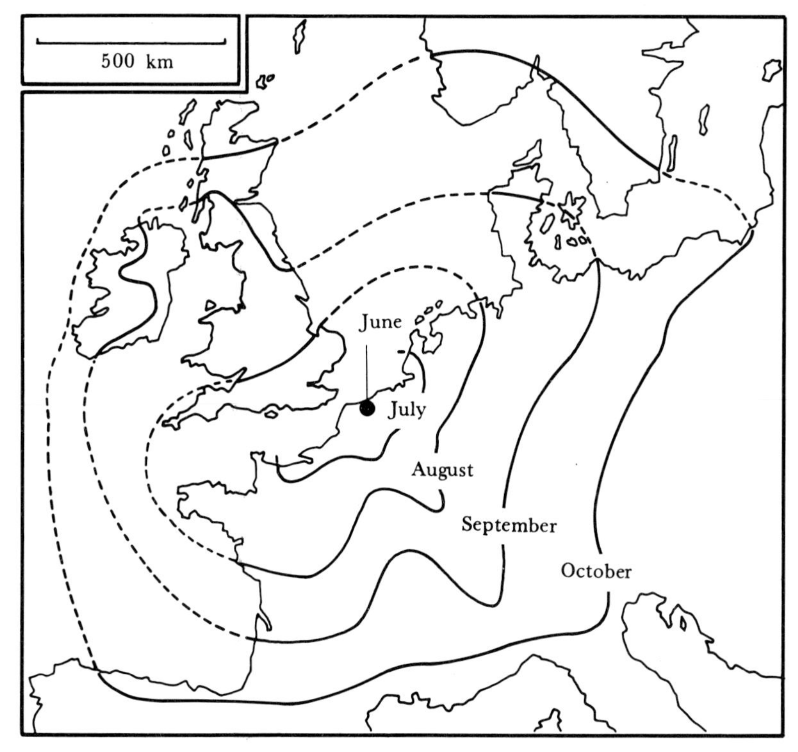

Fig. 3.1 The spread of potato blight through Europe in 1845. The black circle indicates its point of origin in the Low Countries. (After P. M. Austin Bourke, *Nature*, **203,** 805–8. 1964.)

Fundamental to any study of plant pathology, or to the design of control measures, is a knowledge of how much damage is produced by disease. The total effect of the Irish famine was incalculable in both economic and human terms, and it is obvious that even today where a crop disease causes human suffering it is difficult to quantify this when assessing the importance of the disease. The estimation of damage done by diseases is therefore made objectively and purely in terms of crop loss. When such estimates are made the results are difficult to grasp as their scale is so large. For example, European and tropical agriculture may lose 30–50 percent of total crop yields annually through the effects of microbial diseases, among which fungal diseases are probably the most important. When considered in the light of total world agricultural output these percentages represent immense absolute figures.

Information which is more easily assimilable can be obtained from special surveys of the potential dangers of particular diseases in particular regions, but surveys of this kind are usually uncertain, are subject to wide error and the necessary data are only rarely available even in highly-developed countries. Two important surveys, more reliable than most, have been made in the United Kingdom of potential crop losses of potatoes and spring barley due to late blight and barley foliage diseases, respectively. The technique employed in such surveys is to assess the effects of the diseases on crops which have deliberately not been protected against them and in this way to obtain an estimate of the losses that would occur in all crops if no control measures were employed. These hypothetical losses are expressed in terms of 'area wasted', which basically represents that area of the crop that might be used by the disease organism for its growth and which would therefore be lost for production. With respect to *P. infestans* it has been estimated that without suitable protective measures 12 percent of the potato crop, occupying over 20,000 hectares, would be wasted annually through the effects of late blight. Potential losses of barley due to foliage disease amount to over 500,000 hectares, valued at above £60 million (at 1969 cereal prices). These figures are considerable, but refer only to the United Kingdom, and if extrapolations are made to obtain world-wide estimates then they become astronomical in their immensity, yet these are only two of the many staple crops grown as major food sources.

The mechanisms by which plant pathogens cause plant diseases are of great interest since an insight into the relationship between a fungus and its host is essential for the development of effective disease control methods. It is important to remember that the fungus itself is not the disease but that the latter is the manifestation of a number of characteristic disturbances caused by the activity of the fungus on or within the plant. In addition, a single species of fungus can often cause a range of diseases, each with distinctive symptoms, depending on the kind of host it is attacking or on which part of the host is involved. The number and variety of plant diseases is large, as is the number of fungi that cause them, but despite this diversity, a parasitic species can normally be placed in one of two distinct groups. Parasitic fungi are either necrotrophs or biotrophs. This division is not a taxonomic one, so that unrelated species are found together in the same group. Nevertheless, each group has a number of distinct biological characteristics.

Necrotrophs kill host tissues rapidly and then live as saprotrophs on the dead remains. The entire host may be destroyed or, if the activity of the fungus is limited in time or space, only part of it may be lost. The most familiar diseases caused by necrotrophic species are the 'soft rots' of fruits or tubers and these are characterized by the appearance of large areas of watery, often brown, tissues composed of dead cells (Fig. 3.2). Such rots may occur either while the fruit or tuber is still attached to the plant or during post-harvest storage. The fungus normally enters by means of a wound that has been caused by insects or mechanical damage and quickly grows through the tissues. The symptoms of the disease reflect the effect of the fungus on host cells as it advances through them. Its hyphae secrete enzymes and other factors which pass out of them and diffuse into the surrounding host tissues where they affect the physiological and physical integrity of the constituent cells. Since these fungal secretions are extracellular, and diffusible, the pathogen exerts an influence on the host which is effective well beyond the site within the tissues where the fungus is growing. The permeability of host cells is first affected, so that they more easily lose both water and large molecules such as sugars, and their respiration rate rises dramatically as they lose control over their normal metabolic functions. Fungal enzymes that can degrade host cell-wall material then begin to act. Cellulases break down any cellulose present while pectolytic enzymes destroy the pectic cementing materials which hold the cells firmly together, so that cells separate and host tissues become macerated. Host cells are by this time dead, the final result being a typical soft water-soaked lesion. A great variety of root and foliage diseases are closely allied to soft rots of fruits and tubers and all are brought about by means of similar basic mechanisms although the gross results, as expressed in symptoms, are of course not identical.

Fig. 3.2 Apples infected by a necrotrophic, soft-rot fungus *Sclerotinia fructigena*. The fungus is sporulating on the fruit surface. (*Photograph by G. Woods.*)

Wilt diseases, which are common in horticulture, particularly when tomatoes and cucumbers are intensively grown under glass, are also caused by necrotrophic fungi. Here the fungus enters the root system from the soil by way of a wound and moves into the water-conducting elements of the plant. It then grows upwards into the stem within these elements. Its physical activity is confined to these vascular tissues and its presence within them may cause the movement of water from root to shoot to become badly impeded. The foliage therefore wilts as its water supply is reduced, so that it gradually withers and finally dies. Only after the host is dead does the fungus grow out from the vascular elements to invade other tissues and live on them saprotrophically. The remarkable thing about wilt diseases is that symptoms often appear in the foliage very quickly after invasion of the roots has begun, and commonly well before fungal growth in the conducting tissues is sufficient to physically reduce water supply to the aerial parts of the plant by means of mechanical blockage of these tissues. The fungus must, therefore, produce something diffusible which effects this. Although the plugging of vascular elements, when it eventually occurs, is an important factor in the development of the disease, there is evidence that some wilt fungi also produce toxins. These pass to the leaf cells where they in some way act on the cell membranes and drastically alter their permeability characteristics so that control over water loss is removed.

Necrotrophic fungi are not restricted to the parasitic mode of life, and most can live successfully as saprotrophs in the soil or decaying organic matter where they have no difficulty in competing with other saprotrophic fungi. Their well-developed saprotrophic ability is demonstrated by the frequency with which they occur in the absence of host plants, and it has been established that many species are common soil fungi. When in a parasitic situation their behaviour is still fundamentally saprotrophic, the only difference being that they create their own dead organic substrate rather than being dependent on encountering one by chance. Many of them can justifiably be considered to be saprotrophs in the wrong habitat. This is reflected in the lack of ability of most species to enter their hosts by routes other than wounds or other openings. Since they are not specialized parasites they have not evolved mechanisms for breaching intact host surfaces. A consequence of the necrotrophic–saprotrophic habit is that these fungi usually have very wide host ranges. Provided the host can be invaded through a suitable route, a fungus may find that the tissues of a number of different, and unrelated, plant species provide a suitable environment for its saprotrophic growth once they have been killed. *Sclerotium rolfsii*, a soil-borne fungus that causes rapid rotting of host tissues, has been recorded as attacking members of every family of flowering plants.

In contrast, biotrophic fungi have a similar ability to obtain their nutrients from host cells but can obtain them only from living cells. If host cells die, as they eventually must, then the fungus either itself dies, or becomes dormant, or produces spores which carry it to another host plant. Biotrophs have little or no saprotrophic ability under natural conditions. In most diseases caused by biotrophic fungi there is some degree of balance between host and parasite which enables the cells of the former to remain alive for relatively long periods, despite their being exploited by the fungus. The implication is that, in these instances, the host–parasite relationship is an ancient one, the two organisms having evolved together over a geologically long period. This process has also resulted in the existence of narrow host ranges for most biotrophs so that they are, in general, restricted to either a single host species or to a group of very closely related host species. Unlike the necrotrophs they are able to enter their hosts directly through intact surfaces or by means of natural openings, other than wounds, for example through the stomatal pores in leaves.

Diseases caused by biotrophic fungi, particularly rusts, smuts, and mildews of cereals, are some of the most important in agriculture, yet they may cause no massive destruction of plants within a crop. It is easy to visualize the way in which necrotrophs contribute to crop loss but it is more difficult to grasp the subtle effects that biotrophs have on their hosts. While plants attacked by such fungi may not die rapidly, they commonly exhibit symptoms of chronic starvation, and these symptoms are induced by one or more mechanisms that are unique to biotrophs. The simplest situation is, perhaps, that in which a plant is being attacked by a rust fungus. The fungus is restricted to the leaves where it forms distinct, restricted pustules on the leaf surface (Fig. 3.3). Within the leaf tissues, both in and around the pustules, the parasite is in intimate contact with living host cells many of which it penetrates, without killing them, by means of haustoria. Haustoria are modified lateral hyphae that penetrate the host cell wall and invaginate the host's cell membrane. It is through these haustoria that nutrients are absorbed from the host.

A major and early symptom shown by such a diseased plant is that its leaves become chlorotic, i.e. the amount of chlorophyll within them is reduced so that they become pale green or yellow. A result of this chlorosis is that the plant's ability to manufacture carbohydrates through the process of photosynthesis diminishes. Immediately around the rust pustules, however, chlorophyll is retained so that the fungus within each pustule is surrounded by a 'green island' of host tissue rich in carbohydrates and other nutrients (Fig. 3.3). Concomitant

Fig. 3.3 Lower and upper surfaces of coltsfoot leaves infected by a biotrophic rust fungus *Puccinia poarum*. Pustules of limited extent have been formed. The dark margins of those on the left hand leaf are due to the retention of chlorophyll around them forming 'green islands'. (*Photograph by G. Woods.*)

with increasing chlorosis is an increase in the respiration rate of host tissues and, since respiration consumes carbohydrate, this exacerbates the deleterious effects of reduced photosynthesis. In addition, a high proportion of that carbohydrate manufactured by host photosynthesis is transported by the host to those areas of its leaves that are occupied by the parasite, where it then accumulates. This movement and accumulation is under the control of the rust fungus. If only one leaf on a plant is infected, then movement of carbohydrate from the healthy leaves to the infected one takes place, again under the influence of the biotroph. In this way the high nutrient demands of the fungus are satisfied while the bulk of the host tissues, with the exception of those comprising the green islands, are starved of essential carbohydrates because these compounds are continually removed from them. In wheat and other cereals this means that carbohydrate normally destined to be moved to the developing seed is diverted elsewhere, with the result that fewer or smaller grains are produced and there is loss of grain yield.

In rust diseases of cereals the effect of the fungus on crop yield is indirect in that the ears and their grains are not themselves attacked by the biotroph although it interferes with their supply of nutrients. In many other diseases caused by biotrophic fungi there are more direct effects. For instance in smut diseases of cereals the grains are destroyed. Similarly, in downy mildews of brassicas and other plants, where it is the foliage that constitutes the crop, and in some diseases of potato tubers, direct destruction of harvestable tissues occurs. As well as producing local or general starvation symptoms in the host, biotrophic fungi may induce other effects. Many infections lead to the accumulation of unusually high levels of plant hormones (auxins and gibberellins) within host tissues and, since a delicate hormonal balance is essential for normal plant growth, this leads to abnormal growth of the plant. The result is that the host may either grow very tall, and so be weak and spindly, or that it may be extremely deformed. If the fungus is one which invades roots, then the root system may assume a form which renders it inefficient in taking up water and nutrients from the soil. The above-ground parts of the plant consequently wilt or show starvation symptoms.

One of the most sinister aspects of plant pathogenic fungi is the rapidity with which they can spread from one locality to another, and the potential that many have to gradually extend their geographical range or to suddenly appear in countries or continents where they have never been known before. Fungi can be spread in a number of ways, for instance within diseased plant material such as seed, in contaminated soil, or by insects and larger animals. The most common and by far the most important method is through the agency of air-borne spores which are produced abundantly on the surface of diseased plants and are then dispersed in air currents. Most of these spores land relatively close to the site of their formation but a proportion of them remains in the atmosphere and can travel great distances before becoming deposited. Those that are not widely dispersed allow local spread of the fungus within a crop while those that escape from the immediate locality may become responsible for extension of the geographical range of the fungus. The wastage rate for such air-borne spores is high since many will land on non-host surfaces and be lost or, if they arrive on a host, conditions may not be suitable for infection to occur. In addition, only a few survive the rigorous desiccating and radiation conditions to which they are subjected during long-distance travel in the atmosphere. Wastage is compensated for by the enormous numbers of spores that are produced, so that despite the fact that only a minute percentage of viable spores eventually comes into contact with a suitable host, the statistical chances of a successful infection occuring somewhere are quite high.

The explosive spread of a plant disease through the agency of wind-blown spores follows a characteristic pattern. The disease is at first local in extent but then becomes epidemic while being still relatively restricted in area. Within the area of the epidemic the disease becomes more intense and a pandemic is initiated, the fungus suddenly appearing in areas at some distance from the original site of the epidemic. Mention has already been made of the spread of late blight of potato through Europe in the nineteenth century and this still remains the classic example of the epidemic–pandemic pattern of spread. However, this is by no means the only example. Many similarly catastrophic diseases have struck agriculture since then, some of which are even at this moment running their course. Surprisingly, general awareness of these is not widespread. The apparent lack of impact of these diseases on us is attributable to a number of factors. Where edible crops are affected any losses, particularly in well-developed countries, can be made good either from stored reserves or by importation of the commodity from areas which are unaffected. In addition, many disease epidemics or pandemics occur that involve non-edible, but nevertheless commercially valuable, crops which provide luxury products. A compensatory rise in the price of these to offset crop losses tends to go unremarked or to be quickly accepted. Finally, where losses in edible crops occur in underdeveloped countries these may not gain much attention outside these countries, particularly if there is in any case a background of extreme poverty or a history of regular occurrence of other natural disasters. Three of the most interesting recent disease pandemics have involved air-borne diseases of tobacco in Europe, Asia Minor and North Africa; coffee in Brazil; and maize in the United States. Each of these pandemics has spelled out a particular lesson to be remembered by all scientists concerned with crop protection.

'Blue mould' of tobacco is caused by the biotrophic fungus *Peronospora tabacina* which destroys the leaves or, if the disease is not severe, reduces the quality of the crop. During 1957–8 blue mould was found in the United Kingdom on ornamental tobaccos growing under glass. By 1959 it had spread to the commercial tobacco crops in continental Europe through which it rapidly moved eastwards,

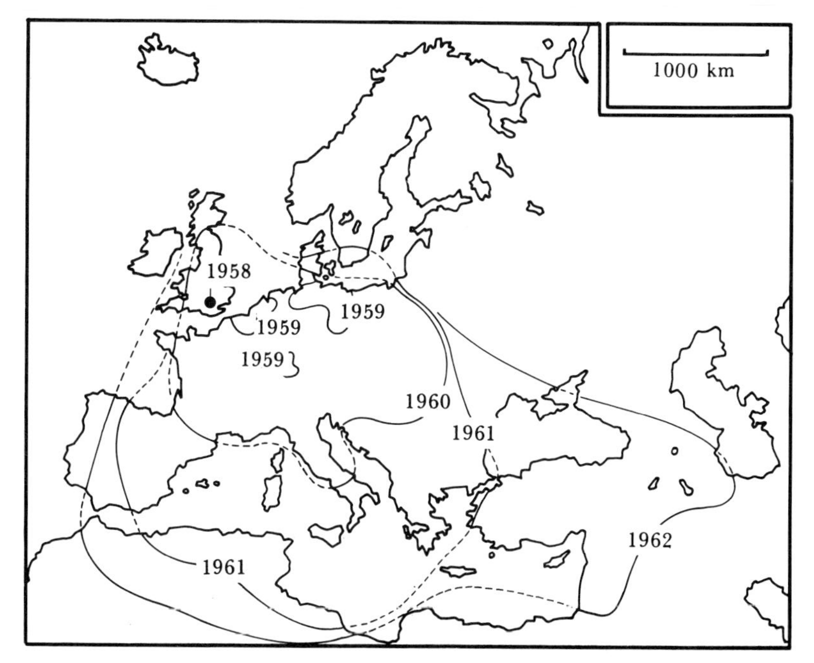

Fig. 3.4 The spread of blue mould of tobacco during the period 1958–62. The black circle indicates its point of origin in the United Kingdom. Slightly modified, according to Corbaz (1964).

reaching North Africa and the Middle East by 1962 (Fig. 3.4). During these years tobacco yields from the field were reduced by almost 50 percent, and over 60 percent of the final cured tobaccos were infected by *P. tabacina*, with a resultant loss in marketable quality. The economic effects were greatest in the poorer countries of Western and Central Europe where the tobacco crop has a relatively high importance. The catastrophic spread of blue mould demonstrates how the development of a seemingly innocuous disease of a non-crop plant can lead to the transfer of that disease to a closely-related, commercially important host.

A rust disease of coffee has played a dramatic economic and social role during the past century which it may be about to re-enact. 'Coffee rust' is caused by a biotrophic fungus *Hemileia vastatrix* which attacks the foliage of coffee trees, causing the leaves to become desiccated and to fall prematurely. When the trees are stripped in this way a new flush of leaves usually appears in the following season and the resulting crop of coffee berries may be normal. However, premature leaf fall year after year exhausts the trees so that they become incapable of producing new wood upon which future crops can be borne. Branches die back and, although the trees are not necessarily killed, they become commercially worthless. Coffee rust was first reported in 1861 in East Africa, where it was affecting wild coffee, and it subsequently appeared in plantations in Sri Lanka in 1869. In a few years the annual coffee yield fell from 42 million kg to less than 3 million kg and by 1890 coffee production had been abandoned, this crop being replaced by tea and rubber. It is almost entirely true to say that it is this disease which transformed the British into a nation of tea drinkers. From 1869 the fungus spread, by means of air-borne spores, eastwards and westwards to reach West Africa by 1918 and the Philippines by 1889 (Fig. 3.5). By the 1960s the disease was established in most coffee-producing areas in Africa. The Western Hemisphere remained disease-free but with inoculum building up in West Africa there was a distinct possibility that wind-borne spores might carry the fungus across the South Atlantic and transmit the disease to the coffee crops of South America. This occurred, and in 1970 the fungus was found in Brazil, where it had probably been present for four or five years before its discovery, and it has since that date been steadily moving southwards and westwards. It seems inevitable that all coffee-growing areas of Latin America will eventually be affected. The disease will not have the same totally catastrophic effect that it had in Sri Lanka in the nineteenth century and it is unlikely that coffee production will be halted. Coffee rust can be controlled with fungicides and varieties of coffee resistant to the fungus are known and can be used. The possible economic consequences of the appearance of *H. vastatrix* in South America are nevertheless serious, particularly since it takes more than ten years for a resistant variety of coffee to successfully replace a susceptible one. It has been estimated that, at present, protective spraying in high risk areas alone would cost the equivalent of 9 percent of the total foreign-exchange earnings of Brazilian coffee exports.

Although coffee rust is not affecting a major food crop, its impact on human populations is potentially just as serious as if it were. This is because of the peculiar position of coffee in the economy of some South American countries. Brazil produces 35 percent of the world's coffee and several South and Central American countries depend heavily on the crop for foreign exchange (Table 3.1). The economic impact of even a low level of rust damage on the economy of such countries

Fig. 3.5 Eastward and westward spread of coffee rust from Sri Lanka during the years 1869–1970.

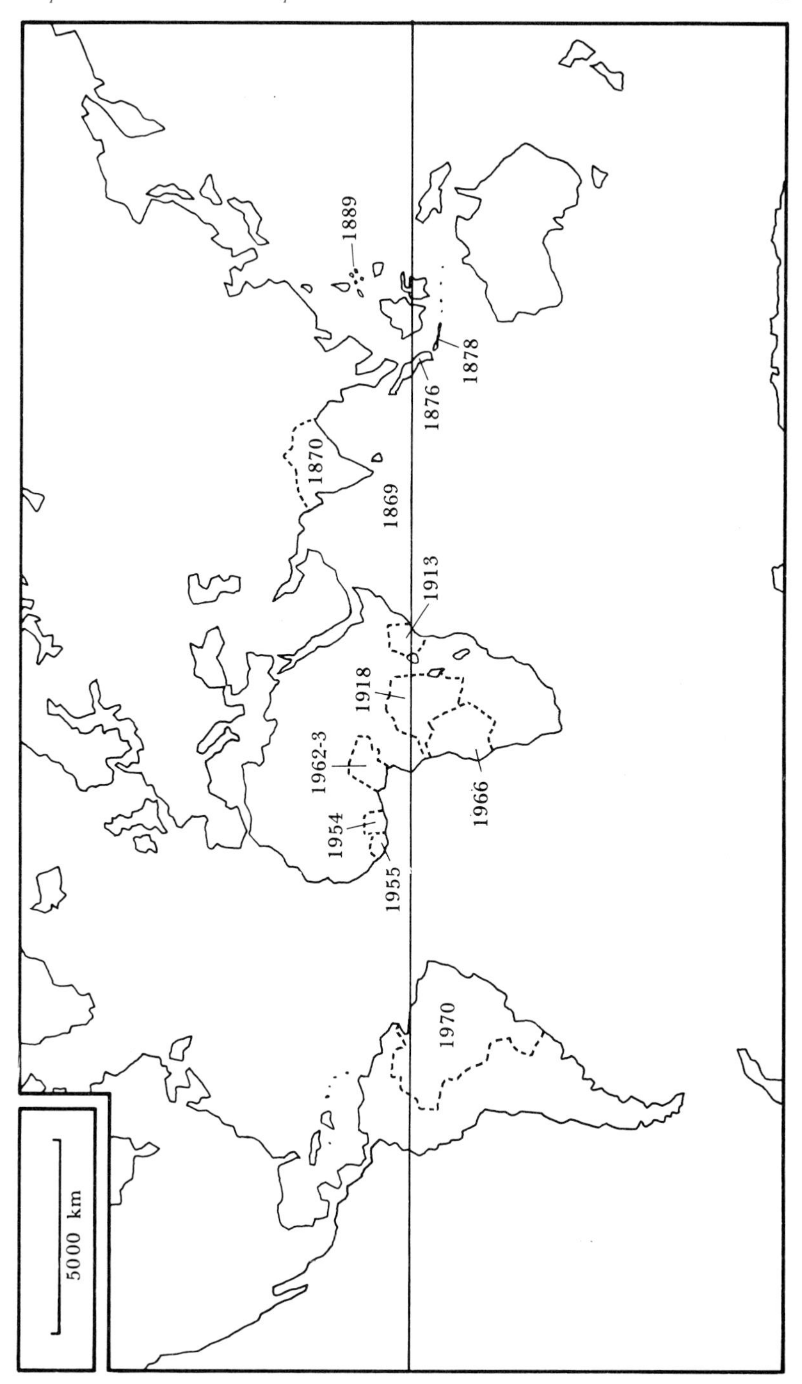
1889
1878
1876
1870
1869
1913
1918
1962-3
1966
1954
1955
1970
5000 km

could therefore be theoretically very severe (Table 3.2). In retrospect, what is remarkable about the South American pandemic is that, although warning notes were sounded as long ago as 1953, no determined efforts were made to prepare for the arrival of the fungus through the development of large-scale growing of rust-resistant varieties of coffee.

The final example to be discussed here of a recent pandemic in a crop is the spread of 'southern corn leaf blight' in the United States. In one way this is the most interesting case since the disease was essentially man-made and resulted from a scientific misjudgement of some magnitude. Maize ranks third in cash value behind wheat and rice among all cultivated crops, and in the United States the land area occupied by it and its total grain yield are greater than for any

Table 3.1 Foreign exchange provided in Latin America by coffee sales based on figures for 1968. (After Schieber, 1972)

Country	Percentage of all foreign exchange
Colombia	68
El Salvador	43
Brazil	41
Haiti	39
Guatemala	33
Costa Rica	31

Table 3.2 Theoretical economic consequences of coffee rust in Central America (Panama to Mexico) based on figures for 1971. (After Schieber, 1972)

Percentage rust damage to crop	Approximate reduction in income (£ millions)	Approximate reduction in hand labour (man-days, millions)
5	9·2	8
10	18·4	15·5
20	36·8	31
30	110·8	46·5

other cultivated crop. Over 80 percent of it is used as livestock feed, the remainder being utilized by industry or manufactured into food for human consumption. One of its great virtues is that it is a relatively healthy crop, supporting fewer pests and being less affected by diseases than any other grain plant. Most of its diseases can be controlled by the development of resistant varieties that are obtained through cross-pollination, and in 1970 maize was considered to be such a clean crop that the United States Department of Agriculture did not think it worthwhile to employ a plant pathologist specializing in maize diseases within the corn belt. This belt comprises the vast area encompassed by the states of Iowa, Illinois, Indiana, Ohio, Minnesota, Nebraska, Missouri, South Dakota, and Wisconsin.

But already by 1970 the seeds of a disaster had been sown. For many years new varieties of maize (Tcms corn) had been coming into widespread use which were derived from male-sterile plants. These varieties were highly susceptible to a then unknown race of a necrotrophic fungus *Helminthosporium maydis*, and all that was required to precipitate disaster were wet weather conditions suitable for the development of the fungus in the crop. These conditions occurred in 1970 and *H. maydis* appeared in Tcms crops, causing extensive rotting of the leaves, stalks, ears and grain. Some crops were totally lost while in other cases yields were reduced by 50 percent. The fungus overwintered successfully in 1970–1 and moved through the crops again in the summer of 1971, although it produced less severe losses than in the previous season. Fortunately, subsequent pandemics were avoided by reverting to the use of non-Tcms varieties that were not highly susceptible to *H. maydis*, but not before serious economic consequences had been felt, not simply at a national level but also on a world-wide scale. The wholesale price of maize rose by over 20 percent and European livestock farmers who depended on it for feed had to turn to barley as an alternative, with the inevitable result that the price of this cereal also rose. It is clear that it is essential to ensure that a cultivated species will never be moulded in such a way that it becomes vulnerable to attack by such a destructive pathogen.

The major part of scientific resources available for plant protection programmes are, quite rightly, absorbed in investigations of the diseases of cultivated crops, primarily food crops. Yet it is important to realize that wild and naturalized plants of all kinds are susceptible to fungal diseases and that, from time to time, epidemics may occur in natural populations. Most of these outbreaks go unremarked, leaving no lasting signs that they have occurred, but others have far reaching, and as yet undetermined, ecological consequences. As is the case with coffee rust, a number of such diseases are at present running their course, their eventual effects being largely a matter of speculation. Two outstanding examples are 'jarrah dieback' of eucalypt forests in Australia and Dutch elm disease in the United Kingdom.

Dieback of jarrah, *Eucalyptus marginata*, was first noted in the dry forests of temperate Western Australia in the early part of this century. Initial symptoms are yellowing of the leaves and the shedding of these without replacement. New, fine branches become either stunted or are lost, this occurring at first in the upper crown of the tree but later extending to the lower, larger branches. Finally, there ensues a phase of extreme debility resulting in death. In seedling trees death occurs within a few days of the appearance of the first symptoms, but in mature trees 15–20 years can elapse before they eventually succumb. Jarrah is the second most important timber species in Australia and it has been estimated that at present in south-western Australia over 80,000 hectares—approximately 5 percent, of the total commercially productive eucalypt forest—has been destroyed (Fig. 3.6).

Jarrah dieback is caused by *Phytophthora cinnamomi*, a soil-borne, phycomycetous species which enters the roots and becomes necrotrophic within them. The consequent rotting impairs the function of the root system, there is a progressive increase in the severity of visible symptoms and, after death of the host, the aboveground parts of the plant are also invaded from the occupied root tissues. *P. cinnamomi* can spread from one root to another or from one tree to another by either root-to-root contact or, when conditions are wet enough, by the production of zoospores which swim through the soil. It can grow very successfully as a saprotroph on dead organic matter and its resting spores can survive in the soil for considerable periods.

If the pathogenic activities of *P. cinnamomi* were confined to *E. marginata* the

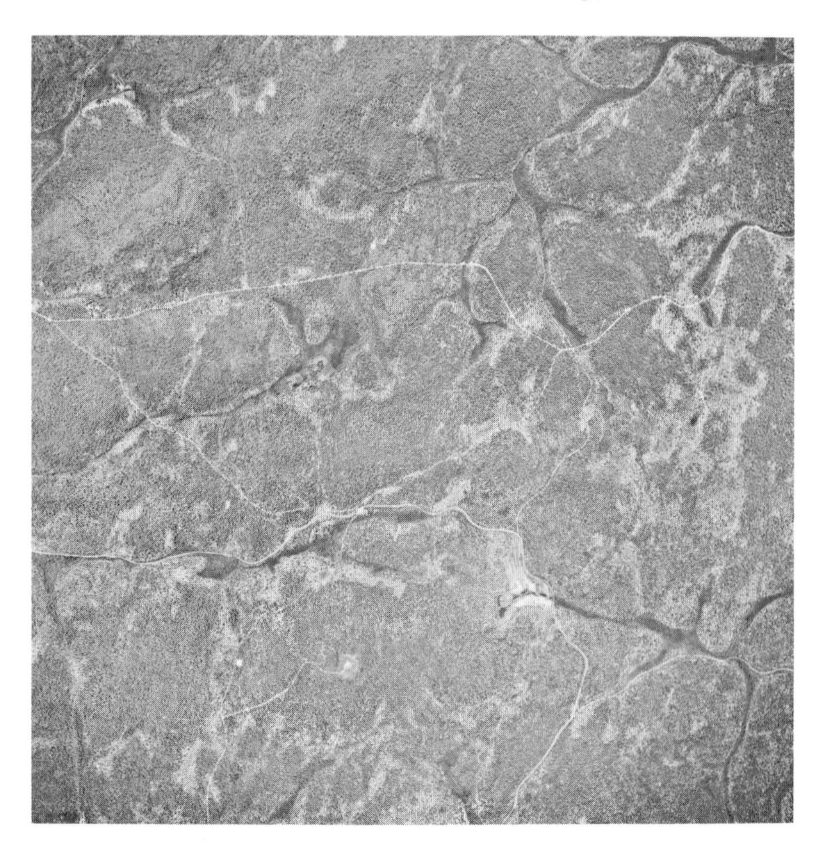

Fig. 3.6 Aerial view from 6,700 metres above jarrah forest in Western Australia. Light patches are areas where trees are dead or dying (*Photograph by Dr F. J. Newhook and Dr F. D. Podger, Forests Department, Western Australia. Copyright Annual Reviews Inc.*)

economic and conservational problems which it is creating would be serious enough. The situation is, however, infinitely more grave than this, and it has gradually become clear that this species is the most destructive fungal pathogen of natural vegetation yet known. It has been found to be capable of causing disease, commonly completely destructive, in over 400 species of Australasian plants belonging to over 130 genera from 48 families. Forest trees other than *E. marginata* are killed and there is often coincident death of the underlying shrub vegetation, so that what was formerly a complex forest system becomes rapidly converted either to a barren community or to a more open community of reduced productivity (Fig. 3.7). As well as severely affecting natural woodlands *P. cinnamomi* can also rapidly destroy the vegetation of heath and swampland. There are well-founded fears that in certain regions of Australia much of the indigenous vegetation, and those animals that are dependent upon it, will be lost. However, this fungus is by no means confined to wild plants but also causes serious losses in shelter-belts and in plantations of pineapple, avocado, and other crops. On the continent of Australia, *P. cinnamomi* is present in Western Australia, South Australia, Victoria, New South Wales and Queensland. It is also well-established in Tasmania and New Zealand. Its rate of spread and the severity of damage caused by it vary widely and depend on complex and little understood interactions between climate, soil conditions and the nature of the species mixture within the vegetation. Thus, the edge of a diseased area may remain almost stationary for a long period or may advance rapidly, at 60–170 m a year, for a period of years. Similarly, in some situations, few trees might be killed or death might be slow or, in contrast, up to 40 percent of mature trees might die in a two- or three-year period.

Phytophthora cinnamomi is soil-borne and is singularly unadapted for aerial dispersal, yet disease is continually breaking out in formerly disease-free localities remote from known sources of infection. Detailed investigations of such outbreaks in the Brisbane Ranges and in Wilson's Promontory National Park (both in the state of Victoria) since 1968 have provided one explanation as to how this occurs. The mapping of newly-diseased areas of forest showed a marked correlation between disease outbreaks and road construction. All infected sites could be traced back to road-making activity. The conclusion was that the fungus was brought into disease-free areas on the tracks of construction vehicles and in road gravel dug from infected soils. In one instance a pile of gravel, less than a cubic metre in volume, deposited at the roadside in 1968 had caused infection of adjacent vegetation by 1972. It was still infective 5 years after being placed there, although it was at that time free of vegetation itself and contained no visible fragments of vegetable matter. It is almost certain that dust and gravel can carry resting spores or dormant mycelium to new localities and that, once established, the fungus then spreads through root contact or zoospore production. Spread along gullies and drainage channels is a common characteristic of the disease.

Since *P. cinnamomi* can survive in soil either as a saprotroph or in a resting phase, has a wide host range and can easily be accidentally disseminated, its control is obviously difficult. Neither the application of fungicidal chemicals to diseased plants nor the burning or mechanical removal of them is likely to prove successful. In addition, the vast scale of the disease defies effective control measures. While it is possible to restrict small, localized outbreaks by injecting the soil around them with fumigants, this method cannot be adopted for treatment of more extensive epidemics. Strict checks on the infectivity of road-making gravel together with suitable quarantine regulations may reduce new outbreaks, but where disease is well established it seems that it must be left to run its course. It is possible, though perhaps unlikely, that the spread of *P. cinnamomi* might, quite spontaneously, lose momentum. What is more probable is that its ecological impact will continue to increase, with long-term effects the nature of which can only be guessed.

Not all speculation concerning *P. cinnamomi* concerns its future and there has been much argument as to whether it is indigenous to Australasia or has been introduced within historical times from another part of the world. Its destructive impact on ancient, and hitherto stable, communities of native plants strongly indicates that it is an alien, and additional circumstantial evidence suggests that it arrived in Australia as recently as the mid-nineteenth century in infected pineapple stocks. Its establishment in New Zealand may date from the early Maori settlements, at which time it could have been introduced with stocks of sweet potato and then subsequently spread by feral pigs, goats and deer.

Fig. 3.7 Jarrah dieback in Western Australia. Large dead trees of *Eucalyptus marginata* with healthy trees and a dense understorey in the background. (*Photograph by Drs F. J. Newhook and F. D. Podger, Forests Department, Western Australia. Copyright Annual Reviews Inc.*)

On a less dramatic scale, over the last decade the spread of Dutch elm disease in England has destroyed large numbers of elms, and in so doing has altered one facet of the countryside. Quite apart from the loss of elm wood to the timber industry, the removal of this important hedgerow and roadside tree has reduced the aesthetic value of many country areas which is unlikely to be restored within a single lifetime.

Dutch elm disease is caused by the ascomycetous species *Ceratocystis ulmi* which in infected trees exists mainly in a yeast-like budding form. These cells are located in the water-conducting elements of the tree and produce a number of toxins which cause yellowing, wilting and, finally, withering of the branches. First symptoms develop during early summer and, if the attack is severe, infected branches may be killed before the end of the same year. However, death of a whole, mature tree within a single season appears to be a rare event. The fungus is carried from tree to tree by beetles which bore and breed within the bark of weakened or moribund elms, including those that are suffering from infections by *C. ulmi*. In England two elm bark beetles are involved, *Scolytus scolytus*, which lives within the bark of the trunk and major branches, and *S. multistriatus* which is mainly restricted to the smaller limbs. *Ceratocystis ulmi* grows within the larval galleries and produces abundant spores contained in droplets of a sticky fluid, so that that when young adults emerge from the breeding tunnels, at any time from May to October, they carry these spores on their bodies. Having emerged, and before breeding again, young adult beetles fly to and feed on young, healthy elm shoots, so that spores of the fungus are introduced into the wounds. In this way the fungus is spread from branch to branch on a single tree or from an infected tree to an uninfected one.

Dutch elm disease was first discovered in France in 1918 but its introduction into Europe, possibly from Asia, was probably much earlier than this. By 1928 it was present and widespread in southern England. In England, damage to the elm population reached its peak in 1936–7 and then declined slowly. The number of trees infected decreased and so too did the general severity of infection, but not before 10–20 percent of the elms in the southern half of England had been destroyed. During that period, and until very recently, control measures involved the felling of dead and badly-diseased trees, less-severely affected trees not being felled as they commonly recovered.

This relatively favourable situation existed until 1965–7 when severe epidemics broke out in North Gloucestershire and South Essex. Since then the epidemic has spread rapidly through England and over 5 million elms have been lost. It has been found that the current epidemic is being caused by a new and aggressive strain of *C. ulmi* and that the recovery rate of moderately-infected elms is relatively low. It has also been convincingly demonstrated that this new strain did not arise from the form of *C. ulmi* already present in England, but that it was introduced in infected timber imported from North America.

Dutch elm disease was first recorded in the United States in 1930, having been introduced in infected timber imported from Europe. It was rapidly disseminated by *S. multistriatus*, which was introduced with the original infected timber, and by a native bark beetle *Hylurgopinus rufipes*, reaching Canada by 1944. For over a century, rock elm, a North American species susceptible to *C. ulmi*, has been imported into England from Canada, mainly for boat building. In Canada this tree is restricted to south-east Ontario which Dutch elm disease had reached by 1950, to become widespread by the early 1960s. During the 1960s the principal ports of entry for rock elm timber were Avonmouth, Liverpool, London and

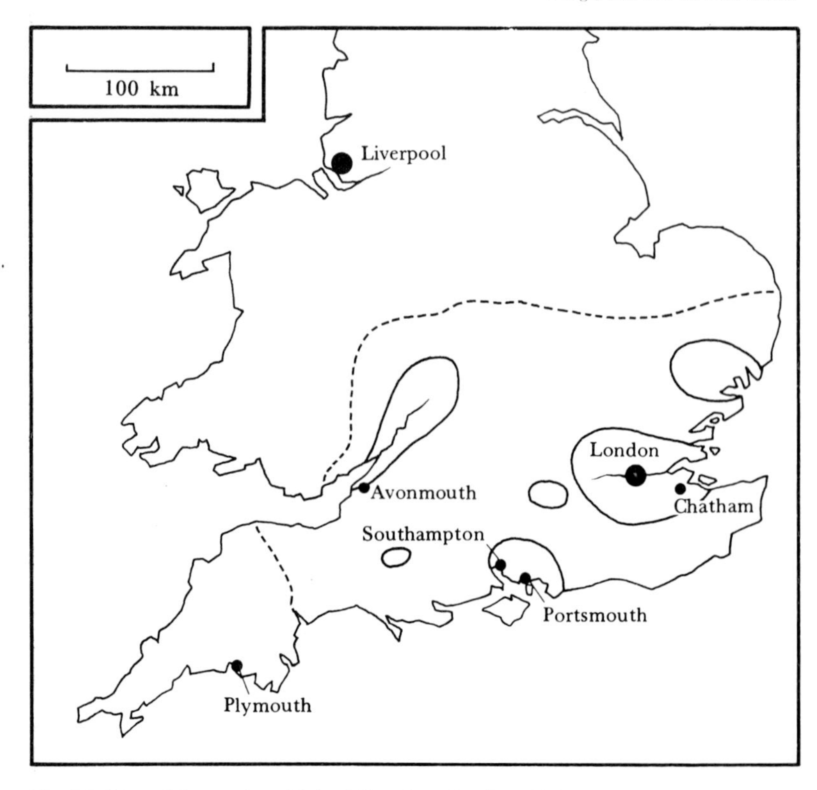

Fig. 3.8 Map of the southern United Kingdom showing chief areas of Dutch elm disease in 1971 (thick lines) and ports of entry and centres of usage of rock elm. The total area surveyed is delimited by broken lines. (After Brasier and Gibbs, 1973.)

Fig. 3.9 Breeding galleries of elm bark beetles in imported rock elm timber. (*Copyright Forestry Commission.*)

Fig. 3.10 Stereoscan electronmicrograph of the North American native elm bark beetle, *Hylurgopinus rufipes*, from imported rock elm timber. (*Copyright Forestry Commission.*)

Southampton, while among the main centres in which it was used for boat building were Chatham, Plymouth and Portsmouth. A 1971 Forestry Commission Survey of Dutch elm disease in southern England strongly indicated a direct association between disease distribution and centres of import or usage of rock elm timber (Fig. 3.8). Subsequent confirmation was obtained in 1973 when infected logs containing live adults, eggs and larvae of *Hylurgopinus rufipes* were intercepted at Southampton (Figs. 3.9 and 3.10). It is ironic that a disease introduced to North America from Europe should return, in an altered and more aggressive form, some 30 years later.

Control measures directed against the new epidemic in England involve using insecticides against the beetle during its free-flying phase, felling severely infected trees and treating less-severely affected trees with fungicides injected under pressure into the main trunk. There is no doubt that these methods, used either singly or in combination, can halt disease in restricted areas and protect individual or small groups of trees that have a particular landscape value. However, whether the epidemic as a whole is affected remains to be seen. The aggressiveness of the introduced strain may gradually decline or the disease may run its course, as it did in the previous epidemic, and some kind of balance may be restored.

The continual occurrence of fungal disease epidemics in crops and in natural plant populations demonstrates, in a very clear manner, that we are never safe from the depredations of plant-parasitic fungi and are unlikely ever to be so. There will be no absolute security, despite our good knowledge of breeding crop plants for disease resistance and our impressive and continually growing armory of fungicidal chemicals.

4 Symbioses and dual organisms

Its slaves are green algae, which it has sought out . . . and compelled into its service. . . . It surrounds them, as a spider its prey . . . but while the spider sucks its prey and leaves it dead, the fungus incites the algae found in its net to more rapid activity. . . .
S. Schwendener, *Verh. schweiz naturf. Ges.*, Vol. 88, 1867

By no means all contacts between fungi and green plants result in disease or death of the latter. During evolutionary time a wide range of partnerships between fungi and plants have developed and the most successful of them are mutual associations which are of great benefit to both partners. In these symbioses the fungus generally obtains all or most of its organic food from the plant in the form of carbohydrates but donates something to the plant in return. During this process some of the plant's cells may be damaged but this occurs at a minimal level and most remain healthy. Like some plant-parasitic fungi, most symbiotic fungi are biotrophic and if their plant partner dies they die too or become inactive.

A characteristic of symbioses involving biotrophs is that the association is long-lived or permanent, and the two partners (the symbionts) are frequently so intimately associated that a distinct, new, living entity is formed. Such 'dual organisms' have morphological and physiological properties not possessed by the partners when they are grown separately, but these properties are not simply the sum of the normal attributes of the fungus and plant added together. Although each to a great extent retains its own characteristics, and these do become combined in the symbiosis, new abilities arise as a consequence of association. Permanent, stable symbioses therefore have morphological and physiological characteristics which together confer ecological versatility on the association, but at the sacrifice of independence of the partners. The versatility of symbiotic systems is of great ecological importance in many situations.

A widely-known example of a well-developed symbiosis is that found in lichens. Lichens are common, conspicuous and can be found growing on rocks, stone walls, buildings and the trunks and branches of trees (Figs. 4.1 and 4.2). Each lichen species has stable characteristics that distinguish it from other lichens and each behaves as a single entity. They are, however, dual organisms resulting from symbioses between fungi and algae. The body, or thallus, of a lichen is usually made up of a single species of fungus and a single species of alga, with the fungus

Fig. 4.1 Thalli of the foliose lichen *Parmelia saxatilis* growing on a stone wall. The largest thallus is approximately 15 cm in diameter. (*Photograph by G. Woods.*)

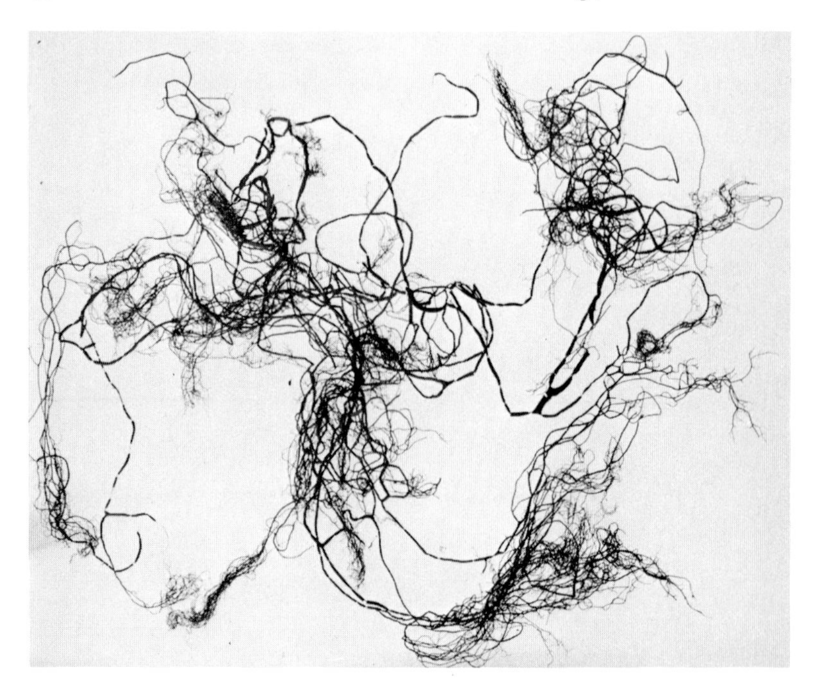

Fig. 4.2 A thallus of the fruticose lichen *Usnea articulata* about 30 cm long found growing on an old tree. (*Photograph by G. Woods.*)

providing the bulk of thallus material. It is the fungus which commonly, but not invariably, determines the form of the thallus, its size, place of growth and rate of growth. Lichens exhibit a great variety of form but there are three basic kinds of thallus (Table 4.1).

Thalli have different degrees of internal organization. At their simplest they consist merely of a loose weft of hyphae within which are held individual algal cells or small groups of cells (Fig. 4.3). The thalli of more highly evolved lichens have a much more complex structure, the algae being located in a distinct layer near the surface. While the greater part of the thallus may be composed of normal but closely-packed hyphae, the tissues above the algal layer may comprise fungal

Table 4.1 Basic forms of lichen thalli

Fruticose	Thallus erect and bush-like or hanging and tassel-like. Attached to the substratum only at its base.
Foliose	Thallus leaf-like or scale-like, creeping horizontally over the substratum and attached to it by root-like threads.
Crustose	Thallus in the form of a crust, closely-attached to the substratum by whole of its under-surface. Sometimes impossible to separate from substratum.

cells that are thick-walled and cemented together to form a tough, protective structure (Fig. 4.3).

The majority of lichens grow very slowly and the margins of thalli commonly advance less than 1 mm in a year. This is due, in part, to the inherently slow growth rates of lichen fungi but may also be imposed by other factors. Lichens normally occupy nutrient-poor substrates and they are rapidly affected by environmental changes, particularly by the amount of water that is available to

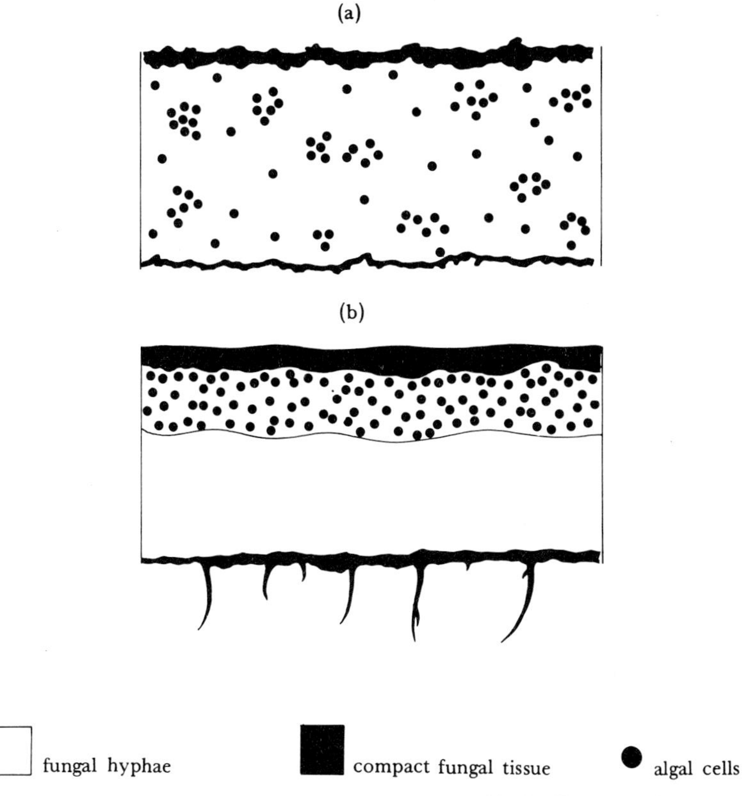

Fig. 4.3 Diagrams of sections through lichen thalli: (a) algal cells scattered; (b) algal cells in a distinct layer.

them. For some lichens, growth has been observed to coincide with the occurrence of measurable rainfall, and their thalli are only capable of growth during such relatively brief, discontinuous periods of water availability. During periods of dehydration, thalli cease growing but do not die, and are highly resistant to the effects of desiccation, reviving quickly when suitable moisture conditions return. Lichens are also extremely resistant to other environmental extremes, particularly when they are in a state of dehydration. Cooling to below −70°C does not impair their viability nor do temperatures as high as 70°C. The ability to withstand water stress and extremes of cold and heat confers a durability on lichen thalli which is reflected in the great age that they commonly achieve.

It has been calculated, from growth rate studies, that some may be up to 4,500 years old.

The exact nature of the lichen symbiosis has been a subject of debate and controversy since they were first discovered to have a dual character in the nineteenth century. Within the last decade, skilful experimental investigations on a number of lichen species have produced a great deal of information concerning the relationships between fungi and algae within the lichen thallus. The background against which these relationships exist is one of obligation on the part of both fungus and alga to associate together. Within the thallus the fungus provides the alga with physical protection by filtering out the harmful, short-wave radiations from sunlight. The fungus also acts as a water-storage reservoir during dry periods, and may insulate the algal cells from extreme diurnal variations in temperature. The carbohydrate needs of the fungus are satisfied entirely by compounds produced by algal photosynthesis, and the fungus to a great extent controls some of the metabolic processes of the alga that are concerned with the manufacture of these carbohydrates.

Free-living algae use the bulk of the carbohydrate which they photosynthesize to produce insoluble structural materials with which to maintain cell growth and division. If algae from lichen thalli are grown in artificial media for long periods they behave in an identical way to free-living algae with respect to this use of photosynthetic carbohydrate. Within the thallus, however, lichen algae are in some way prevented by the fungus from producing insoluble carbohydrates, and instead manufacture mainly soluble compounds which leak from their cells and are absorbed by the fungus. In this way growth and division of the algal cells is severely restricted and this restriction may be contributed to, in part, by a physical restraint imposed by the enveloping fungal tissues. However, photosynthetic activity of the algal cells in unimpaired so that they are continually producing quantities of carbohydrate which are surplus to their requirements. In this way the fungus obtains the carbon compounds with it needs and at the same time maintains the alga, alive and well, within the lichen thallus. Algal growth is, of course, not entirely prevented, otherwise thalli would not be capable of growth.

In addition to requiring carbohydrates, lichen fungi, in common with most other fungi, require vitamins and nitrogen compounds. It has been shown that lichen algae can synthesize the vitamins required by their fungal partners and that these are leaked to the fungus through the algal cell wall. Lichen fungi can obtain their nitrogen in the form of soluble nitrate or ammonium salts from the substrate on which they are growing, but in certain circumstances the alga may supply it. Many lichens contain species of a peculiar group of algae, the Cyanophyceae or 'blue-green algae', which are remarkable in being able to build up complex nitrogen compounds by absorbing and chemically fixing atmospheric nitrogen. They can thus manufacture within their cells all the nitrogen-containing materials that are required to maintain growth in the absence of any nitrogen other than that contained in the air. In such lichens, restriction of algal growth by the fungus leads to the release of surplus soluble nitrogen compounds which are then absorbed by the latter.

The durability and peculiar physiology of lichens confer upon them the kind of versatility which allows them to colonize borderline habitats. These are ecological niches where temperature conditions and the poor availability of nutrients or water exclude most higher plants and even mosses. This means that in certain geographical situations lichens may constitute the major vegetation type, and for this reason they are of ecological importance. For example, because of low temperatures and frozen subsoils, over very large areas of Arctic tundra the dominant ground cover is composed, not of plants, but of foliose lichens which may be from 10–25 cm high and form an almost continuous carpet. In areas where bare rock predominates, lichens may contribute significantly to soil formation through the erosion and fragmentation of the mineral surfaces upon which they are growing, although this process is obviously a very slow one. Lichens growing in regions which lack other forms of vegetation, or developing on nutrient-poor substrates, also provide food and cover for insects and other invertebrates that would otherwise not be able to survive.

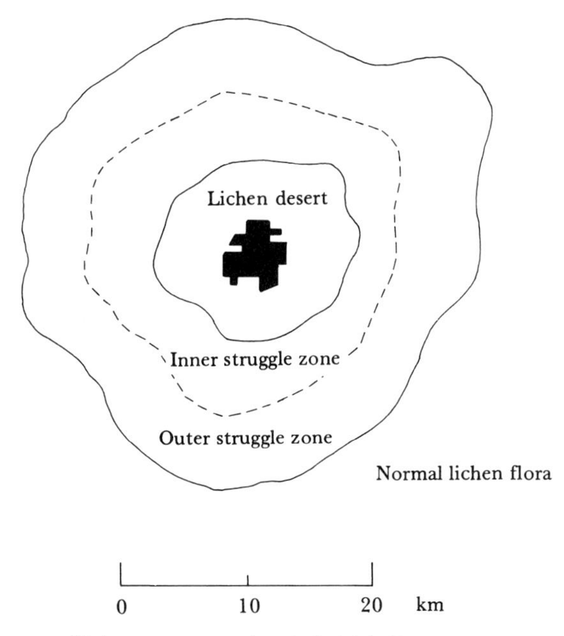

Fig. 4.4 Diagram of lichen zones around an industrial city.

Under some conditions they also provide useful forage for mammals. In the Arctic, herds of reindeer and caribou supplement their winter diet with lichens and, under very severe conditions, lichens can constitute up to 95 percent of their total food intake. Although lichens are low in protein, calcium and phosphorus, they are rich in carbohydrate and may be important for the survival of such animals in bad winters. In some areas of British Columbia, lichens growing on trees are the only winter food available in quantity to caribou, and thalli suitable for grazing are present in amounts of up to 3,000 kg per hectare. Much of this material is on branches that are out of the caribou's reach but a considerable quantity, over 300 kg per hectare, grows at a height of 3–5 m. This too, under normal circumstances, would be out of reach but the heavy winter snowfall, over 2 m deep, brings it within range of the animals. In addition, frequent winter tree-falls bring down more lichen which is then grazed from the fallen timber. Lichen fragments also continually drop from standing trees during the winter and become hidden in the snow. In spring the melting of the snow reveals them and they can provide early feed at rates of 21–125 kg per hectare.

One form of environmental stress which most lichens cannot survive is that imposed on them by domestic and industrial atmospheric pollution. Since the latter part of the nineteenth century it has been noted that lichens are commonly absent, or poorly developed, within and around industrial cities and that this is associated with environmental pollution: lichens which are brought in from country districts invariably die. By mapping the pattern of lichen distribution around large cities in both Europe and America it has been shown that the general features of their distribution are always the same (Fig. 4.4). There is first a central zone, the 'lichen desert', from which lichens are almost entirely absent. This is surrounded by a belt, the 'struggle zone', within which lichens occur but in a stunted or poorly developed condition. Relatively few species may be found in the struggle zone but those few that are present may be abundant. The struggle zone can often be subdivided into an inner and outer region. Within the inner region foliose lichens predominate, while in the outer one it is fruticose forms that are in the majority. The outer part of the struggle zone grades into the surrounding area where lichen numbers and growth are normal. The distance from the centre of the lichen desert to normal regions varies, of course, with the size of the city and its climate, but can be greater than 15 km. Similar patterns of zonation are found around isolated industrial plants, particularly those involved with smelting and refining. There is often an extension of the lichen desert over areas of open country downwind of the source of pollutants (Fig. 4.5).

Many different chemical compounds contribute to atmospheric pollution but a major factor, within cities at least, is the level of sulphur dioxide in the atmosphere. This chemical is emitted by industry on a vast scale. Global estimates indicate emission rates in the region of 132 million tons a year, with the United States alone contributing over 30 million tons of this. The bulk of sulphur dioxide released to the atmosphere results from the combustion of hydrocarbon fuels and, in particular, coal. Lichens are especially sensitive to sulphur dioxide and most are killed by levels in excess of 0·015 parts per million in the atmosphere. This is a level at which most higher plants are unaffected. Sulphur dioxide acts on lichens in two ways. First, its presence makes the thallus environment acid and this discourages growth. Secondly, it is directly toxic and causes rapid breakdown of the chlorophyll in the algal cells so that they can no longer carry out photosynthesis. This bleaching of chlorophyll is only reversible if the damage caused is not too severe and if favourable atmospheric conditions return quickly.

Lichens are sensitive to even low levels of sulphur dioxide because they accumulate sulphur within their thalli. For instance, thalli of *Usnea subfloridana* removed from an unpolluted habitat and exposed to a polluted atmosphere near a large industrial city (Newcastle upon Tyne, UK) were found to contain 1,010 parts per million of sulphur after two months. Rapid accumulation of chemicals from dilute sources normally allows lichens to efficiently colonize nutrient-poor

Fig. 4.5 Map of north-east England to show the distribution of the lichen *Parmelia saxatilis*. Black circles indicate the inner limits of thalli growing on exposed rock and industrial areas are stippled. The wind stars indicate the average number of days in a year on which the wind blows in the particular direction shown. (After Gilbert, 1970.)

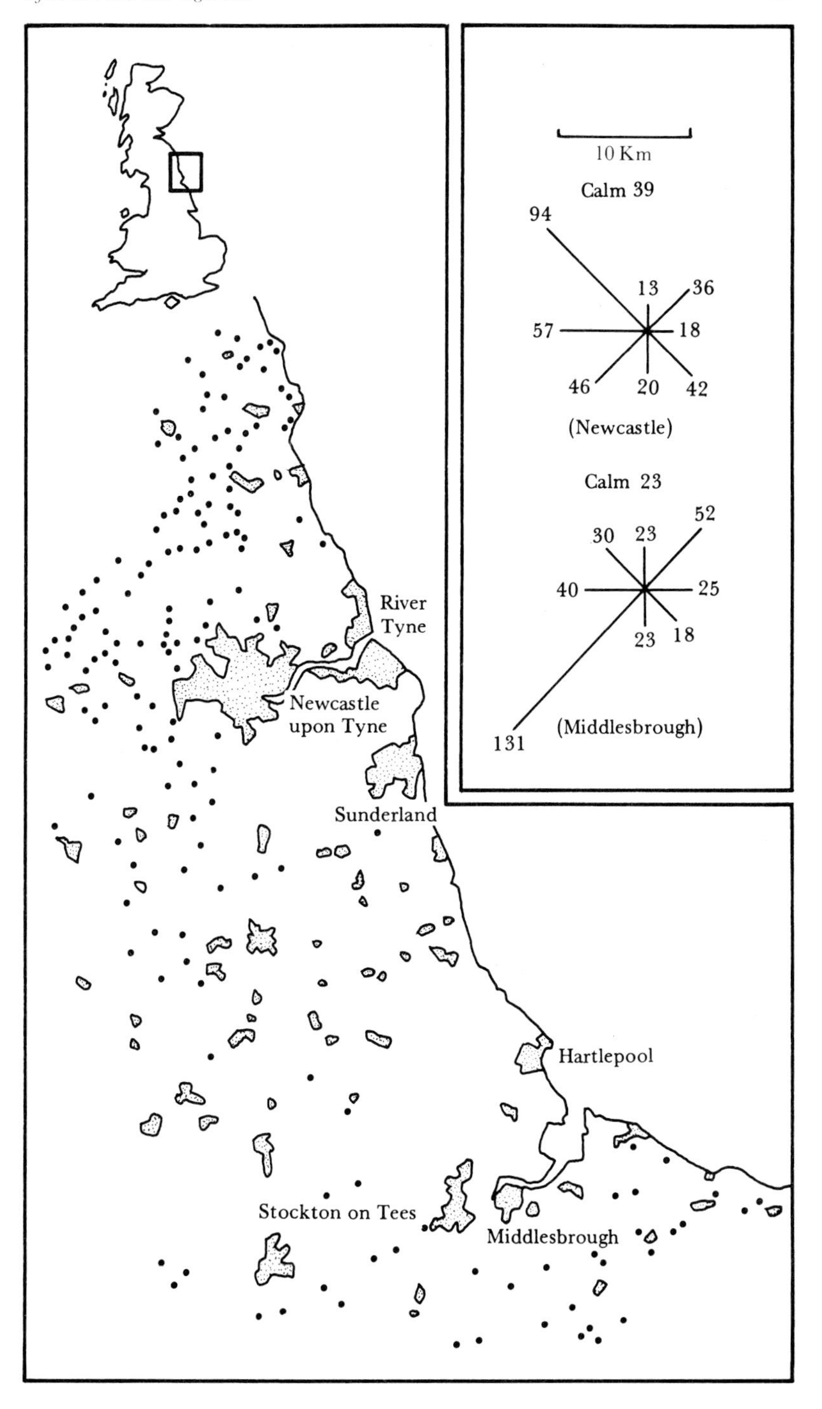
10 Km
Calm 39
94
13
36
57
18
46
20
42
(Newcastle)
Calm 23
52
30
23
40
25
23
18
131
(Middlesbrough)
River Tyne
Newcastle upon Tyne
Sunderland
Hartlepool
Stockton on Tees
Middlesbrough

habitats, but in situations where there is pollution the same physiological property is responsible for their destruction. The sensitivity of lichens is such that, as a group, they are reliable indicators of atmospheric pollution. By noting the size of thalli the number of species present and the kind of substrate on which they are growing at any particular site, it is possible to estimate very accurately the annual average levels of sulphur dioxide at that site (Fig. 4.6).

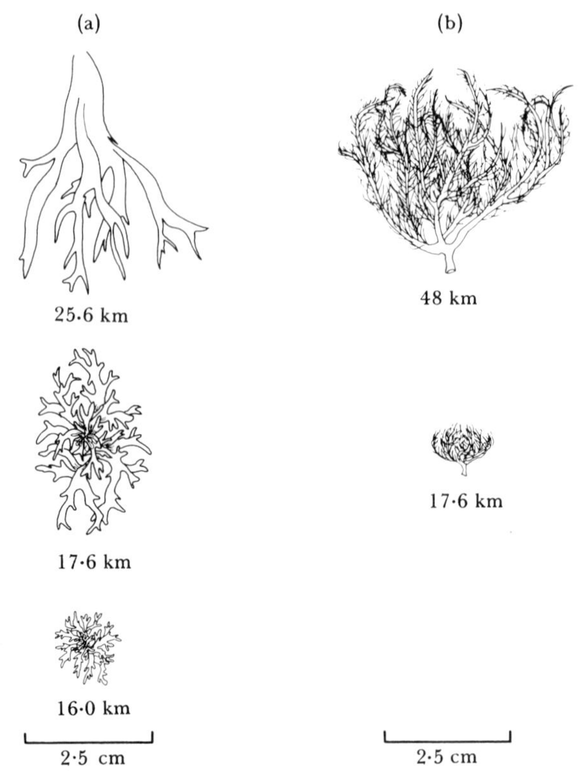

Figure 4.6 The effect of pollution on growth of two fruticose lichens: (a) *Evernia prunastri*; (b) *Usnea subfloridana*. The distances from the centre of Newcastle upon Tyne, UK, at which the thalli were collected are shown. (After Gilbert, 1970.)

There are, however, toxitolerant or toxiphilous lichens which thrive in areas of high pollution. Toxitolerant species are successful because not only are they resistant to the damaging effects of sulphur dioxide, but they also benefit from the lack of competition from non-tolerant species which are severely damaged. Toxiphilous lichens are actually stimulated in some way by certain pollutants. It is not always an easy matter to determine whether the presence and spread of a lichen is due to toxitolerance or to toxiphily. For example, *Lecanora conizaeoides* is a species that was rare in the United Kingdom until the middle of the nineteenth century, but which has since become abundant in and around cities, spreading apparently against the background of rising industrial pollution.

It is remarkably resistant to sulphur dioxide under acid conditions but it has not yet been established whether it might, in addition, have peculiar nutrient requirements which are satisfied by urban pollution.

Possibly less familiar kinds of symbiosis are those that exist between biotrophic fungi and the roots of higher plants. These fungus–root associations are called mycorrhizas and, although they are for the most part hidden from us, they have great ecological importance, and even economic and agricultural significance. It is probably the rule rather than the exception that the roots of plants growing under natural conditions develop close, mutual partnerships with fungi, and that the successful establishment and subsequent vigorous growth of many major groups of plants, including crop species, is favoured by mycorrhiza formation.

A mycorrhiza is initiated when a plant root becomes infected with an appropriate soil-borne symbiotic fungus. The subsequent development of the fungus upon and within the root leads to the formation of an organ which is part fungus, part plant. Physically, the characteristics of mycorrhizas are very diverse, there being at the one extreme pronounced modification of the infected root, while at the other there may be little external evidence that the root is infected. On the basis of their morphological and anatomical features, three broad kinds of mycorrhiza can be distinguished. First, there are ectomycorrhizas, in which the root is entirely surrounded by a well-developed, compact sheath of fungal mycelium, from the interior of which hyphae arise that pass into the root, growing between its cells but not penetrating them. Root hairs, which are responsible for absorbing soil nutrients, are lacking in mycorrhizal roots. Secondly, there are endomycorrhizas, which differ from ectomycorrhizas in that the mycelium external to the root is relatively poorly developed and is not aggregated into a sheath. Hyphae pass into the root but, as well as growing between host cells, they also penetrate them. Finally, there are ectendomycorrhizas that are in some ways intermediate between ecto- and endo-mycorrhizas and which possess some of the features of both of them. A sheath may be present, although it may not always be well developed, while the hyphae within the root penetrate its cells as well as passing between them. While not minimizing the importance of these intermediate forms, it should be pointed out that the greater part of knowledge of mycorrhizal function has been obtained from studies of ecto- or endomycorrihizas.

Approximately 3 percent of the total number of plant species have ectomycorrhizas, the majority of such plants being deciduous or evergreen forest trees (Fig. 4.7). Their fungi are in the main basidiomycetous species of such common genera as *Amanita*, *Boletus* and *Tricholoma*, and the fruit bodies of these arise from mycelia that are ultimately connected to the roots of the trees with which they are symbiotic. Ectomycorrhizal fungi can be divided into three ecological groups depending on the degree of dependence on their host. First, there are species that are normally free-living saprotrophs but which are also capable of forming mycorrhizas with suitable hosts. The stinkhorn, *Phallus impudicus*, is an example of such fungi. Then there are species that are normally mycorrhizal with a wide range of hosts but which can also survive as saprotrophs. The common earthball, *Scleroderma aurantium*, is one of this group. Finally, there are the majority of known ectomycorrhizal species which have little or no ability to live unassociated with the roots of their hosts. These obligately ectomycorrhizal fungi for the most part have a wide host range but some have extremely restricted ranges; *Boletus elegans* for instance is confined to species of larch.

In contrast to free-living, saprotrophic fungi, obligately ectomycorrhizal fungi have a reduced ability to decompose leaf litter and other forms of soil organic

Fig. 4.7 Ectomycorrhiza of the European beech. The short, stubby lateral roots are enveloped in a fungal sheath and lack root hairs. (*Photograph by G. Woods.*)

matter. They can utilize the simple and relatively readily available carbohydrates that are present in virgin organic debris, but do not seem to be able to produce in sufficient quantity the various enzymes which are necessary for the efficient exploitation of more complex carbon compounds. A consequence of this is that they are adversely affected by competition from more efficient, litter-inhabiting fungi and depend largely, if not entirely, on their hosts for a supply of assimilable carbon sources. Soluble carbohydrates produced from host photosynthesis pass from root cells into those hyphae that are growing between them, and from these to the sheath where they accumulate in the form of fungal storage compounds. Where fruit bodies are being formed, carbon compounds are presumably moved to these from the sheath by means of the hyphae which connect them.

The growth of tree seedlings in enhanced by the formation of ectomycorrhizas and this beneficial effect is greatest when the plants are growing in nutrient-poor soils. The degree of infection of root systems is often most intense in such soils and it is clear that symbiosis increases the ability of the root system to take up nutrients, even though root hairs, the normal absorbing structures of the roots, are absent from ectomycorrhizas. Phosphorus, nitrogen, potassium, calcium and other soil nutrients are absorbed by the outer layers of the sheath, move to the hyphae lying between the root cells and are then passed from these hyphae to the root tissues and hence to the rest of the plant. The uptake of mineral nutrients from the soil by mycorrhizal roots takes place at 2 to 5 times the rate found in comparable non-mycorrhizal roots. Ectomycorrhizas are, therefore, extremely efficient in obtaining minerals from soils within which non-mycorrhizal roots would grow very poorly. For example, mycorrhizal Red Pine trees grow well on exposed glacial outwash, an extremely nutrient-poor soil, and their growth rates are similar to those achieved when they are grown on nutritionally adequate sandy soils.

The general outcome of this kind of symbiosis quite clearly leads to ecological success for both fungus and tree and allows both organisms to exploit habitats in which neither alone might grow successfully. Similar benefits accrue to both partners in a number of endomycorrhizal associations, but the manner in which these endomycorrhizas function differs in a number of ways from that of ectomycorrhizas. Two widespread and important kinds of endomycorrhiza—ericaceous mycorrhiza and vesicular-arbuscular mycorrhiza—have only recently become reasonably clearly understood. Both of them are formed by fungi which probably have some capacity for a free-living, saprotrophic existence in the soil, and once within the roots the fungi usually penetrate host cells. However, even when invasion takes place on a large scale there is minimal damage to the cells, and this appears to be partly due to the ability of the host's cytoplasm to digest the penetrating hyphae.

The Ericaceae is a flowering plant family which contains a number of common shrubs, the most familiar being the heaths and heathers, bilberry, cranberry and rhododendron. The roots of these plants terminate in a series of very fine, mycorrhizal, absorbing rootlets that lack root hairs. The endomycorrhizal fungus forms a loose weft of mycelium on the rootlet surface from which hyphae penetrate host cells, forming prominent hyphal masses or coils within them (Fig. 4.8). After a time, these penetrating hyphae are digested and disappear. When fungi from ericaceous mycorrhizas are grown in pure culture, they are invariably found to be dark, slow-growing forms that do not fruit readily, and so cannot be firmly identified, but which are probably ascomycetous.

Many ericaceous plants are typically inhabitants of very acid soils of low nutrient status, for instance peaty moorland soils. Vast areas of the Northern Hemisphere are covered by such soils, yet these support good growth of heaths, heathers and various berried shrubs, and it is probable that this obvious ecological success is a direct result of their mycorrhizal condition. It has been demonstrated that mycorrhizal seedlings of the common ling or heather, *Calluna vulgaris*, and of the Large Cranberry, *Vaccinium macrocarpon*, grow significantly faster than non-mycorrhizal seedlings of the same age (Fig. 4.9). In addition, the nitrogen content of symbiotic seedlings may be over twice that of their non-symbiotic counterparts. The explanation of this increased growth and dramatically enhanced nitrogen status seems to be that mycorrhizal roots are capable of absorbing soil nitrogen compounds, possibly amino acids, which are unavailable to non-mycorrhizal roots. Humus-rich soils that support typical ericaceous vegetation contain amino acids which could be taken up by the external weft of hyphae on the root surface. Nitrogen compounds would then be transported to the interior of the root within hyphae and would ultimately be released into the cytoplasm of host cells upon digestion of the hyphal coils and branches.

While it is clear that ericaceous plants benefit from mycorrhizal infection through improved nitrogen nutrition, the advantages accruing to the fungus from association with the root are not wholly apparent. Evidence suggests that endomycorrhizal species have a very restricted capacity to utilize cellulose so that, as is the case with ectomycorrhizal fungi, they may to some extent be dependent on their hosts for a supply of carbon compounds. Considerable movement

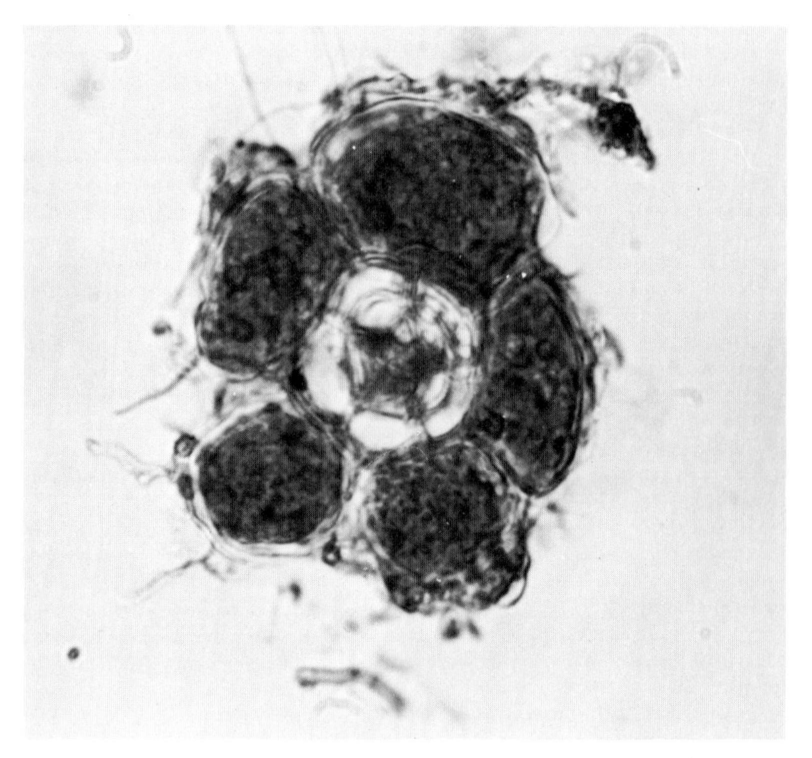

Fig. 4.8 Transverse section of a hair root of heather, *Calluna vulgaris*, showing its endomycorrhizal fungus within the outer cells. Coiled hyphae within these are being digested by the host. (*Material prepared by Dr D. J. Read, photograph by G. Woods.*)

of photosynthetic compounds from ericaceous host to fungus has been demonstrated, but the degree to which this is of ecological advantage to the fungus is uncertain. The widespread occurrence of fungi capable of forming mycorrhizas with ericaceous plants, together with their presence in soils from which ericaceous hosts are normally absent, suggests that, despite their inability to utilize cellulose, they do have a capacity for a saprotrophic existence. Until more is discovered about their mode of life in the soil this aspect of the symbiosis will remain a matter of debate.

It is obvious that the maintenance of particular vegetation patterns over large areas of humus-rich, nutrient-poor soils depends on the efficient functioning of ericaceous mycorrhizas. However, there are other endomycorrhizas that are possibly more ecologically important on a geographical scale and have a present, and also possibly a future, importance in agriculture. These are the vesicular-arbuscular mycorrhizas which, unlike ectomycorrhizas and ericaceous mycorrhizas, are formed by a small group of aseptate, phycomycetous fungi. It seems probable that the majority of mycorrhizal plants in the world flora have this kind of association, and vesicular-arbuscular mycorrhizas are not restricted to any particular plant group. In the tropics and the Southern Hemisphere, where ectomycorrhizas are not common, they are the typical mycorrhizas of forest trees.

As in ericaceous mycorrhizas, vesicular-arbuscular mycorrhizas have a loose, external mycelium that extends into the soil and from which the penetration of host root cells takes place. Within penetrated cells, the hyphae either become

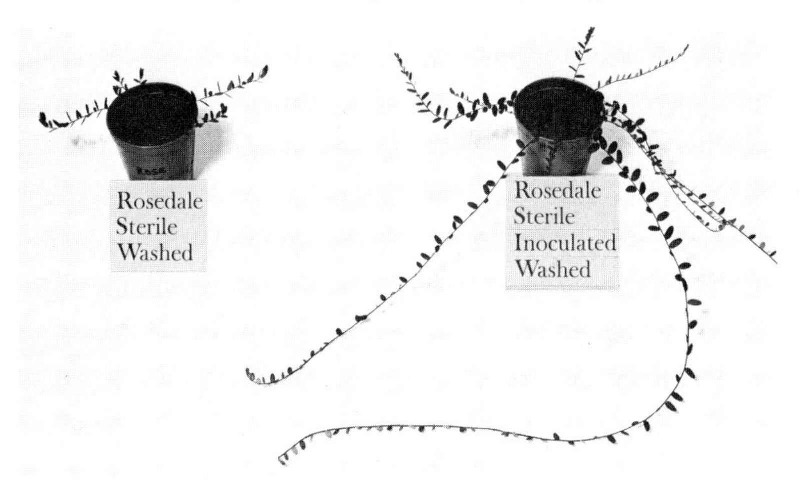

Fig. 4.9 Cranberry plants, *Vaccinium macrocarpon*, grown in nutrient-poor acid soil with an endomycorrhizal fungus (right) and without the fungus (left). (*Material prepared by Dr D. J. Read, photograph by G. Woods.*)

coiled or, more typically, branch profusely to form bush-like arbuscules (Fig. 4.10). In addition to these structures large, oil-rich vesicles may be produced on both the penetrating hyphae and on those hyphae growing between host cells. Arbuscules normally disintegrate quickly, possibly being digested by the host cytoplasm, and their contents are released into the host cell. Externally, infected roots appear to be quite normal and retain their root hairs.

Vesicular-arbuscular fungi seem to belong to a single phycomycetous family the Endogonaceae, but despite numerous attempts none of them have been cultured away from their host roots. Direct observation of soil has shown that these fungi can grow upon and spread from dead organic material in soil but, since pure cultures have never yet been obtained, no detailed information on their saprotrophic ability and capacity for a free-living existence is available. Nor is it clear how much, if any, of their carbon requirements are met by the movement of photosynthetic compounds from the host. Despite these difficulties with regard to the fungi, the consequences for the host plant of infection by vesicular-arbuscular species are at least partially understood.

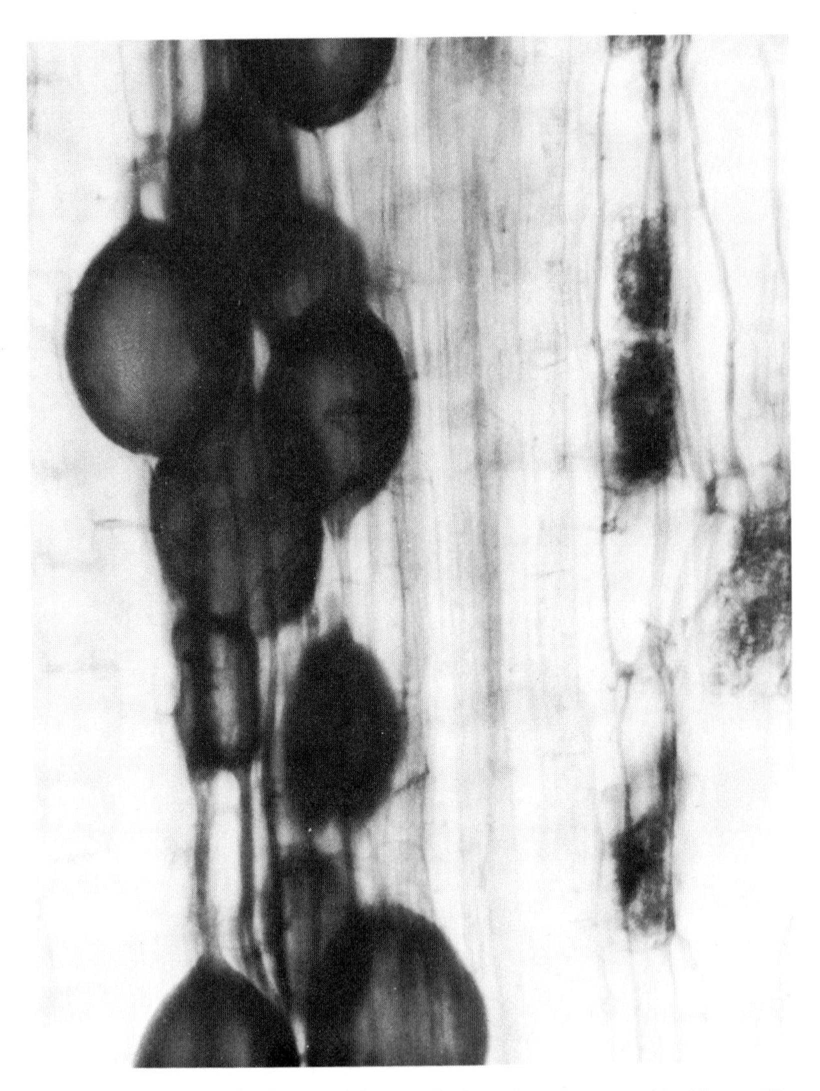

Fig. 4.10 Root tissues of maize containing a vesicular-arbuscular mycorrhizal fungus. The dark contents of host cells to the right of the prominent, ovoid vesicles indicate that arbuscules are being digested within them. (*Material prepared by Dr D. S. Hayman, photograph by G. Woods.*)

Vesicular-arbuscular fungi enhance host growth and this is most marked on phosphorus-deficient soils, although the amount of increased growth is very variable. There is ample evidence that this effect is due primarily to the increased movement of phosphorus into the plant, and it appears that the fungus acts as a bridge for the movment of phosphorus compounds from the soil to the interior of the host root in the same way that ericaceous mycorrhizas facilitate the passage of nitrogen compounds. There is the additional possibility that, not only does the fungus take up and move phosphorus compounds to the host, but it also in some way increases the ability of the roots themselves to do so. However, if infected roots are cleared of their symbiotic fungus by treatment with fungitoxicants their phosphorus uptake is reduced to the same level as that found in non-mycorrhizal roots.

Given that the external mycelium of vesicular-arbuscular fungi can provide an efficiently distributed surface for the absorption of phosphorus compounds, two important features of the mycorrhiza should be remembered. First, the infected root retains its root hairs, which presumably continue to function normally. Secondly, there is frequently only scanty development of external hyphae, and even when these do develop in abundance, entry points into the root may still be relatively few in number. Surprisingly, numerous external hyphae and frequent entry points are not necessary for significant increases in phosphorus uptake to occur. For example, it has been calculated that four hyphal connections per millimetre of root, with unbranched hyphae extending from these into the surrounding soil to a distance of 20 mm, would allow a 10 to 60-fold increase in phosphorus uptake by that root.

Vesicular-arbuscular mycorrhizas are not restricted to wild plants but are also characteristic of a wide range of crop species, including cereals and legumes. Clearly, these symbioses must contribute to the well-being and high yields of many crops but to what extent has not yet been quantified. It is interesting to speculate that in the future varieties of crop plants may be developed with the ability to form endomycorrhizas to a degree where dependence on expensive mineral fertilizers for economic yields will be greatly reduced. Perhaps as well as breeding for resistance to fungal diseases, selection for susceptibility to mycorrhizal infection might prove to be a worthwhile process.

Fungi that are parasitic upon or mutualistically associated with plants have a series of obvious and profound effects on both man's crops and on populations of wild, ecologically important species. Less obvious, and certainly less well-known, are the direct interactions of fungi with animals, including man. These too are at the one extreme destructive while at the other they are essential for the ecological success of certain animal groups.

5 Ailments of man and his livestock

The head is often diseased with an ill that children have oft ... and we calleth the ill Tinea, a moth, for it fretteth and gnaweth our parts of the skin of the head as a moth fretteth cloth.

John de Trevisa, *Bartholemeus de proprietatibus rerum*, 1398

During their lives a large proportion of individuals will become infected, on one or a number of occasions, by fungi. The most common kinds of infection are those of the skin and although these can be irritating and even disfiguring they are rarely, if ever, fatal. In other, less frequent, infections deeper body tissues are involved so that severe debilitation or death may result. Deaths from fungal infections in the United States account for between 0·02–0·03 percent of deaths from all causes, and while this figure appears to be small it is twice that for the combined fatalities due to scarlet fever, typhoid, whooping cough, diphtheria, dysentery and malaria. As well as causing diseases directly, many fungi, while being incapable of invading body tissues, can induce severe allergic responses if their spores are inhaled, and there is clear evidence that many kinds of hay fever and asthmas are caused by fungal spores and not by plant pollen.

While many diseases caused by bacteria and viruses are decreasing, the incidence of some that are caused by fungi is either maintaining a constant numerical ratio in proportion to population increase and general mortality rates or is increasing. This is due to a combination of two factors. First, there has recently been an increase in awareness that fungi are important disease organisms in humans and this has resulted in improved clinical procedures. Symptoms resulting from fungal infection that might formerly have been attributed to other causes are now more often being properly diagnosed. Secondly, there has been a real increase in fungal infections due largely, and paradoxically, to advances in the treatment of other diseases. The widespread use of broad spectrum antibiotics, the utilization of immunosuppressive drugs after surgery and the administration of steroids can all lead to an increase in the susceptibility of the body to attack by fungi. In addition, the life span of geriatric patients, who would otherwise succumb to such diseases as cancer, is now commonly prolonged by means of intensive care. Both the age of the patient and the kind of treatment used often favour secondary fungal infections.

The human body can provide a favourable habitat for fungi if the latter are able to gain access to its tissue, are not then subsequently excluded by the body's defence reactions and are capable of growth at body temperature, which is normally 37°C. Body tissues contain abundant water, have high carbohydrate levels and possess easily-available nitrogen compounds in the form of amino acids and proteins, so that the nutritional requirements of most fungi are easily satisfied. In general, the fungi involved in diseases of man fall into one of two distinct categories, being either exogenous or endogenous species.

Characteristically, exogenous fungi are able to exist in a free-living state in nature where they are very successful saprotrophs that inhabit a wide range of organic substrates. Most are extremely common fungi. Having entered the body they do not find conditions there inimical to their growth and proceed

Table 5.1 Some exogenous fungi that cause disease in man

		Tissues involved	**Habitat when free-living**
Phycomycetes	*Coccidioides*	Respiratory tract, lungs, viscera, bones	Desert soils
	Mucor and *Rhizopus*	Lung and gastric tissues	Dung, rotting fruit
Ascomycetes	*Histoplasma*	Lymphatic system, lungs, spleen, liver, kidney, nervous system	Bat and bird dung, soil
	Neurospora	Corneal tissue of the eye	Soil
Basidiomycetes	*Cryptococcus*	Lung tissue	Pigeon droppings
Fungi Imperfecti	*Aspergillus*	Lung tissues, nasal sinuses	Decaying organic matter
	Fusarium	Corneal tissue of the eye	Soil
	Cladosporium	Skin and the underlying tissues	Soil and wood

to colonize the tissues. While doing so they still behave as saprotrophs, rapidly using the tissues as organic substrates, and only rarely have they evolved parasitic mechanisms for exploiting the host. They are thus unspecialized fungi which are growing out of place. That the human body is not their natural abode is indicated by the observation that diseases caused by them are very rarely transmitted between infected hosts, but are almost always contracted through contact with soil or other substrates in which the fungi are growing. A great many species of free-living saprotrophs must therefore be potentially capable of producing disease in humans, but those most commonly encountered are relatively few in number, being for the most part Fungi Imperfecti, together with some yeasts, Phycomycetes and Basidiomycetes (Table 5.1).

Some infections may be symptomless, or are characterized by mild symptoms which quickly disappear, but generally three broad kinds of serious disease are caused by exogenous fungi. First, infection can be superficial with fungal activity limited to the skin and hair, or to the non-vascularized tissues of the eye. Secondly, mycetomas may be formed. These are localized infections under the skin and are characterized by suppuration and swelling. Finally, disease may take the form of a mycosis. Mycoses are infections that involve deep tissues, typically those of the vital organs, and are potentially systemic, i.e. not restricted to one part of the body. Exogenous fungi enter the body through wounds or by means of inhaled spores, and growth of them proceeds either through hyphal extension or by the building of yeast-like cells. Hyphal growth leads to a slow spread of the fungus through the tissues and the maintenance of a compact mycelial colony. However, many species that are hyphal outside the body assume a yeast form when growing within it as a response to the high nutrient levels

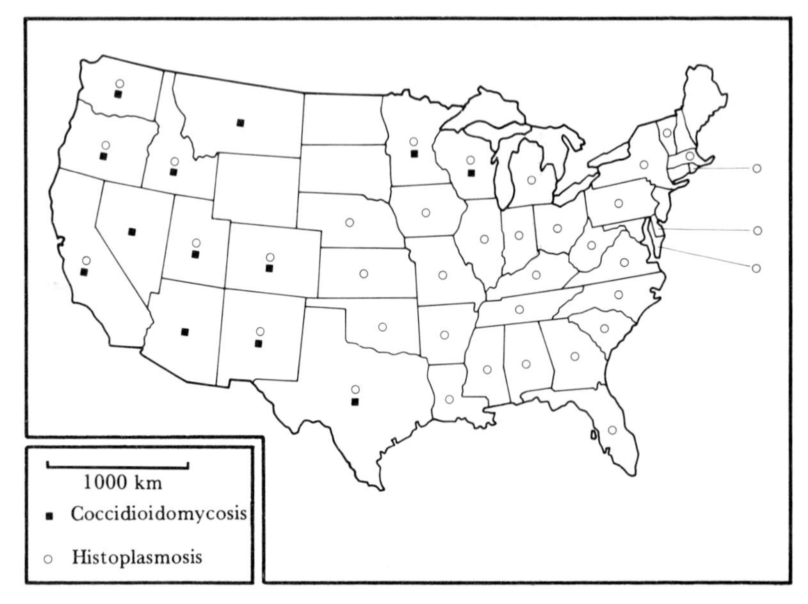

Fig. 5.1 The pattern of incidence of Coccidioidomycosis and Histoplasmosis in the United States based on hospital cases. (Based on data from Hammerman, Powell and Tosh, 1974.)

and relatively high temperature which exist in the host tissues. Fission or budding of the fungal cells allows rapid dissemination of the pathogen throughout the body. This frequently takes place via the blood stream after white blood corpuscles have engulfed the alien cells in an attempt to destroy them. The fungal cells survive within the corpuscles and are released at new sites within the body when the white cells subsequently die and rupture.

Some of the most common and widespread diseases that are caused by exogenous fungi involve the lungs, although the fungus may move from these organs to other parts of the body as the disease progresses. Probably the best studied of such diseases is coccidioidomycosis, which is caused by the phycomycetous species *Coccidioides immitis*. This fungus inhabits desert soils in geographical areas characterized by mean July and January temperatures of 27°C and 7°C, respectively, and an annual rainfall of 10–50 cm occurring mainly during a short winter period (Fig. 5.1). The fungus and the disease which it causes appear to be mainly restricted to the New World in the desert areas of California, Arizona, New Mexico, South-Western Texas, Mexico, Central America and Venezuela, and the Chaco of Argentina and Paraguay. *Coccidioides immitis* is not evenly distributed throughout the soil but occurs most frequently in and close to the burrows of desert rodents where conditions of temperature and humidity favour its growth and survival.

As the fungus grows it produces large numbers of small spores which are readily disseminated in air currents, and infection occurs when these are inhaled (Fig. 5.2). On reaching the lungs the spores rapidly develop into large spheres which then divide to produce further spores. Growth and spore production cause the formation of lesions in the lung tissues, and spores may spread to other tissues by way of the blood stream. Further lesions may thus develop in the skin, the

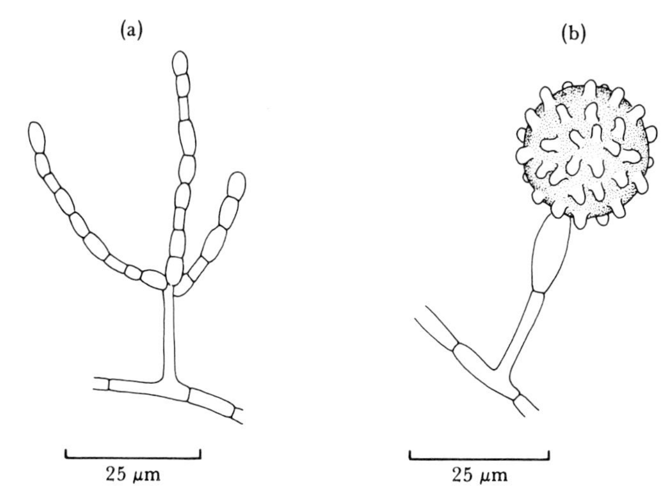

Fig. 5.2 Fungi harmful to man: (a) conidia of *Coccidioides immitis*; (b) a single conidium of *Histoplasma capsulatum*.

bones (commonly including the skull), joints and visceral organs. Finally, the central nervous system may become affected with fatal results.

No ethnic group is absolutely immune to infection, but among members of long-established, stable populations living within the life zone of the fungus only mild influenza-like symptoms are produced, usually in childhood, which quickly disappear. It is interesting to note that desert rodents also become infected but are apparently not badly affected. The disease can be more severe in individuals from non-indigenous human populations. Symptoms include high fever, painful breathing, headaches, rapid weight loss, pains in the leg joints and severe meningitis. About 40 percent of such infections result in noticeable symptoms but only 0·5 percent develop a severe, progressive mycosis. With treatment, fatality rates are approximately 0·25 percent.

Within the United States the large-scale post-war migrations to the south-western States have resulted in the widespread occurrence of coccidioidomycosis within these new populations. It has been estimated that in California and Arizona there may be between 5–35,000 active infections annually. This estimate does not take into account infections that result in very mild symptoms, so that the absolute figures may be much higher. In certain districts within these States almost the whole population has, at some time, been infected and 20 percent of them have developed an illness severe enough to require medical care. Sometimes coccidioidomycosis appears well outside its normal geographical range and has, for example, been occasionally diagnosed in Europe. In these instances infection has occurred from spores borne on fruit, grain, or packing materials originating from areas where the disease is endemic, but no further spread takes place because affected persons are not infectious and the fungus cannot grow in non-desert conditions. It is perhaps fortunate that the distribution of *C. immitis* is so limited, since there is neither a way to control the growth of the fungus in the soil nor any practicable method of preventing inhalation of the spores. Some reduction in the risk of infection can be obtained within some population centres by seeding bare ground with grass, or hard-surfacing dirt roads so that the amount of contaminated dust is diminished, but carrying out these measures is clearly not feasible on a wide scale.

Another major infectious disease due to an exogenous fungus is histoplasmosis which is endemic in the United States and also occurs around the world in both temperate and tropical zones (Fig. 5.1). The causal organism is an Ascomycete, *Histoplasma capsulatum*, and it is estimated that 20 percent of the population of the United States have been exposed to and infected by the fungus. As with *Coccidioides*, air-borne spores are involved (Fig. 5.2) and similarly the great majority of infections are so mild as to be unnoticed. Inhaled spores give rise to yeast-like cells which show preference for the tissues of the lungs but can also infect the spleen, liver, kidneys and bone marrow. The yeast cells quickly divide and at the same time host cells multiply rapidly, so that normal tissues are replaced by disorganized cell masses. The mycosis may become systemic, and severe symptoms include fever, anaemia, weight loss and multiple ulceration. This form of infection is almost invariably fatal.

A curious feature of histoplasmosis is that it tends to have a village or rural distribution and this is related to the occurrence of suitable habitats which are capable of supporting growth and sporulation of *H. capsulatum*. The fungus only grows in soil that is enriched by the droppings of birds or bats and is common in and around poultry houses and within caves. The disease may be contracted when bird-pens are cleaned or demolished and spore-laden dust is inhaled, or when caves inhabited by bats are visited. Exposure in urban districts may also take place if dust originating from the soil beneath large starling roosts is inhaled.

A wide range of antibiotic compounds is available for the treatment of diseases caused by bacteria and viruses, but successful therapy of severe mycoses is rather more difficult, there being available relatively few effective antifungal compounds that are non-toxic to the patient. One of the simplest compounds is potassium iodide, but although it is often beneficial it must be administered in large doses and has unpleasant side-effects. The only well-proven antifungal antibiotic is amphotericin B which rapidly enters the fungal cells and causes them to quickly respire and reduce their internal reserves of carbohydrate. They also leak essential metabolites so that they literally become starved to death.

The great majority of human fungal diseases are caused by *endogenous fungi* and these organisms, unlike exogenous fungi, do not usually exist in a free-living state in nature. They are commonly members of the characteristic fungal flora of the body, living either on the skin or on the mucous membranes, and the diseases which they generate are normally limited to these tissues. The transmission of disease takes place through direct or indirect contact between infected and uninfected individuals and, again in contrast to exogenous fungi, endogenous species are therefore specialized, the body providing their natural habitat. Endogenous fungi fall into two groups. The first is a large one comprising the dermatophytes or 'ringworm fungi' which attack keratin-rich tissues, namely the skin, hair, or nails, causing diseases called dermatophytoses. The second group is small, being made up of yeast-like fungi that normally occupy the mucus of the membranes of the mouth, alimentary canal, rectum and urinogenital system. The presence of mucosal yeasts is a normal condition of the human body but some of them can cause disease if the physiological status of the host changes in such a way that its tissues become susceptible to invasion.

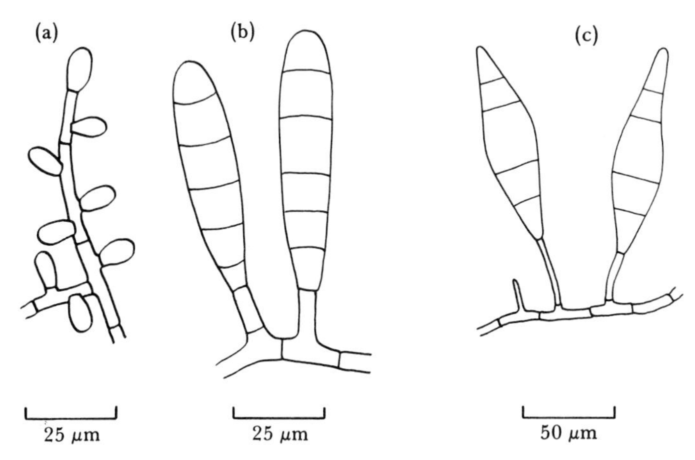

Fig. 5.3 Fungi harmful to man; (a), (b) microconidia and macroconidia of *Trichophyton*; (c) macroconidia of *Microsporum*.

Most dermatophytes belong to two closely-related genera of the Fungi Imperfecti, *Microsporum* and *Trichophyton*, although other fungi can also cause dermatophytoses (Fig. 5.3). It is interesting to note that many species that inhabit such keratin-rich substrates as feathers or keratin-enriched soil belong to the same two genera. However, dermatophytes are rarely isolated from these habitats and free-living keratinophilic fungi are not usually involved in dermatophytoses. These two groups of fungi are taxonomically close and both utilize keratin but they are biologically distinct from one another.

Although dermatophytes occasionally invade deep tissues, it is usually the skin and its appendages that are colonized and the fungi grow saprotrophically upon these, digesting the keratin and making use of other nutrients contained within keratinized structures. As hyphae grow and branch throughout the keratinized tissues the latter become disrupted, and this leads to the appearance of the typical symptoms of ringworm diseases. Hair and hair follicles may be destroyed and the outer layers of the skin become raised, peeling when wet and flaking when dry. Inflammation and weeping of the affected areas commonly occur. No part of the body is resistant to dermatophytes but they appear most frequently on the scalp and feet where they produce the complaints termed respectively 'tinea capitis' and 'tinea pedis'.

Tinea pedis, or 'athlete's foot', is an infection that is familiar to most people and is caused by *Trichophyton mentagrophytes* or, less commonly, by *T. rubrum*. The fungus inhabits the skin of the feet but in particular the webs between the toes. Here the warm, humid conditions, together with the normal continual maceration of the interdigital skin, favour its development. Tinea pedis is relatively infrequent in women, whose footwear usually permits a good degree of air circulation around the feet, and in individuals who habitually go barefoot. *Trichophyton mentagrophytes* is spread from infected to uninfected persons by means of shed fragments of skin containing mycelium. These may contaminate floors, furniture or clothing and subsequently be picked up on the feet. If suitable conditions for the development of the fungus prevail, then disease becomes apparent after a short incubation period. Athlete's foot is most common in situations where large numbers of people are regularly exposed to skin fragments that have been shed from infected feet. For example, surveys of colliery workers making regular use of pithead baths have shown that over half can have clinical symptoms of fungal infection, with 21 percent of these cases being attributable to *T. mentagrophytes*. This compares with infection levels of 4–8 percent in groups not making use of the baths. Disease incidence is also high among users of public swimming baths and in groups of children and adults that are institutionalized. The transmission of other ringworm fungi can occur between both wild and domestic animals and man, and animals can act as important reservoirs of skin diseases. There is, however, at least one recorded instance of transmission in the reverse direction. A dog with ringworm symptoms was examined and *T. rubrum* was isolated from its diseased skin. This fungus very rarely causes disease in animals but in this instance the dog's owner, who had athlete's foot, was for some reason in the habit of rubbing the dog's back with his infected feet. Normal hygiene can prevent infection by most ringworm fungi but even if they become established on the body they can usually be quickly eradicated by the topical application of fungicidal compounds. An effective antifungal antibiotic, griseofulvin, can also be used.

Among the mucosal yeasts the most frequently encountered are species of *Candida* and one, *C. albicans*, is ubiquitous in man on the membranes of the mouth, digestive tract and vagina. It may also live unobtrusively on the skin. *Candida albicans* cannot usually invade the mucous membranes or their underlying tissues, but predisposition of the body to infection can result from a range of factors. Endocrine gland disorders and the use of antibiotics, steroids or immunosuppressive drugs can all result in predisposition. *Candida albicans* is capable of causing a range of diseases, called candidoses, which may either be restricted to the superficial body layers or involve deep tissues. If invasion is deep, then all organs of the body are susceptible and the result is frequently fatal. For tissue invasion to take place the yeast cells must give rise to hyphae, since only the latter are capable of bringing to bear the necessary mechanical force to penetrate through and between host cells. The changes which take place in the host that predispose tissues to invasion are thought to directly induce the filamentous phase of fungal growth, but the fundamental reasons why this transition occurs are not known.

While *C. albicans* is one of the few endogenous fungi that cause serious, potentially, fatal, disease the majority of infections due to it are relatively minor, for example some forms of vaginitis are candidoses. Another candidosis, 'thrush', which occurs in new-born children, seems to be part of the natural order of things.

During the final months of pregnancy the population of *C. albicans* within the vagina increases, perhaps as a result of changes in the hormonal balance of the body. When the full-term foetus passes out of the uterus, fungal cells may find their way to the mucous membrane of the child's mouth and they then invade the superficial tissues causing white, raised lesions. This candidosis is easily controlled and may in any case regress spontaneously.

Fungal diseases of domestic and wild animals are at least as common as those of man and are probably much more frequent. While most of these diseases are unimportant a number of them have a social or economic significance far greater than is generally recognized. Mention has already been made of animals acting as infective reservoirs of endogenous fungi, and this is particularly true of domestic pets. Dogs may harbour *Microsporum canis* which can cause ringworm of the body and scalp in man, particularly in infants, and the incidence of such a disease in cities is often related to the number of dogs that are kept as pets. Perhaps more seriously, some diseases of domestic animals are commercially important, and there are generally two levels at which they make impact. First, diseases may result in financial loss, often considerable, but cannot be said to be of economic significance in the broad sense. For example, ringworm of such expensive animals as greyhounds and race horses may temporarily prevent them from competing or disfigure them so badly that they are unusable or unsaleable. Secondly, fungal diseases may regularly, and on a moderately large scale, contribute to the loss of crop animals, thereby having a significant and negative effect on the agricultural economy. Two diseases of this kind, both caused by exogenous fungi, are mycotic abortion in cattle and aspergillosis in birds. Bovine abortion is not uncommon, and in approximately 6 percent of cases the foetuses are found to have been attacked by fungi, particularly by *Aspergillus fumigatus*, a free-living saprotrophic species. How the fungus gains entry to the uterus is not clear. *Aspergillus fumigatus* and other *Aspergillus* species can also cause a lung disease, aspergillosis, if their conidia are inhaled. This disease frequently occurs among birds, including ducks, chickens and turkeys. Modern techniques of rearing these birds often create an ideal situation for development of the fungi and the consequent production of massive numbers of their spores. Conditions within rearing units are warm and humid and there is abundant organic matter in the form of droppings mixed with litter. This, together with the crowding of large numbers of birds in a confined space, can lead to severe outbreaks of disease and not inconsiderable losses.

Fungal diseases among wild animal populations assume importance when the animals involved provide a food crop which is regularly harvested. Many molluscs, crustacea and fish belong to this category of animals, and outbreaks of fungal disease among them can reach epidemic proportions. One such disease which is currently spreading through Europe is the freshwater 'crayfish plague' caused by the phycomycetous species *Aphanomyces astaci*. This attacks the tissues of the joints or those between the body segments and rapidly kills the animal, possibly through the production of a toxin which affects the nervous system. The disease first appeared in Italy in 1860 and spread north and east to Sweden and Eastern Europe where it is currently established and still spreading. Since no races of resistant crayfish apparently exist it has been said that the total extinction of the animal is eventually not impossible.

Herring provide a cheap source of protein and many local industries are based on this fish. It has for some time been recognized that fungal disease can bring about mass mortalities in herring populations. *Ichthyophonus hoferi*, a fungus which is found in both marine and freshwater environments, can cause systemic mycoses of fish, attacking the heart and other organs and rapidly destroying them. In acute infections herring die within 30 days of infection. Documentation of the occurrence of the disease is poor but in the western North Atlantic alone, in the Gulfs of St Lawrence and Maine, there have been six major recorded outbreaks since 1898 spaced 14–25 years apart. The most recent of these was in 1954–5 when an estimated 50 percent of the herring population was destroyed,

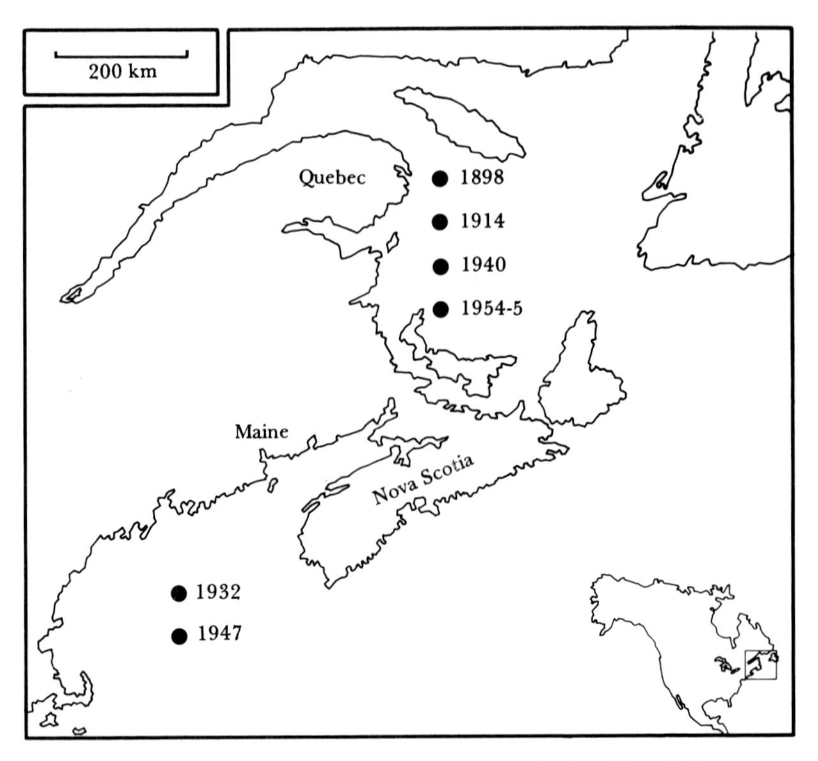

Fig. 5.4 Approximate periods during which large-scale outbreaks of fungal disease have occurred among herring in the Gulfs of St Lawrence and Maine since 1898. (Based on data from Sindermann, 1970.)

producing a decline in landings to about half their previous level (Figs. 5.4 and 5.5). Such large decreases in herring populations are important but are not permanent, since the herring has a high reproductive potential, and there is a gradual return to its former population size. In some instances the loss of herring is compensated for by an increase in other fish. For example, in the case of the 1954–5 outbreak, cod, which are not susceptible to the fungus, fed on disabled herring to such an extent that cod landings doubled during the years immediately following the outbreak of the disease.

Recently, interest has been aroused in diseases of fish species that are normally not considered to be economically important but which, on consideration, are far from being unimportant. During 1966 severe outbreaks of disease appeared in Irish and English rivers among salmon, trout and coarse fish. Mortality rates were so high that legislation was introduced to control the transport of live fish and eggs out of affected areas. Loss of fish on a large scale could result in the extinction of commercial salmon and trout fishing, upon which some rural communities depend. In addition, in those areas where a lucrative tourist industry based on game fishing is a major source of employment, loss of fish would be serious.

'Salmon disease' has been unknown since 1748 when a 'fatal distemper of roach in captivity' was described in a letter to the Royal Society of London. Outbreaks

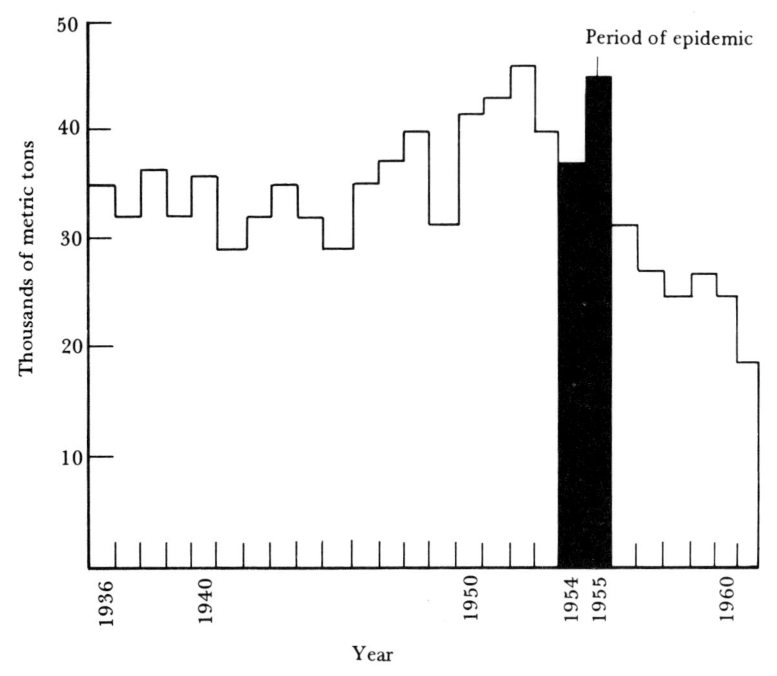

Fig. 5.5 Herring landings from the Gulf of St Lawrence 1936–61 showing the effects of the fungal disease epidemic of 1954–5. (Based on data from Sindermann, 1970.)

sufficiently serious to arouse government interest occurred in Welsh and Scottish rivers in 1877 but the cause was not determined although fungi were strongly implicated. The disease, now referred to as ulcerative dermal necrosis (UDN), appears first as greyish patches on the body surface. An obvious mycelium develops on these areas and bleeding ulcers are then formed. The resulting lesions may be so deep that the muscles are exposed and hyphae may penetrate more than 1 cm into this muscle tissue. Infected fish continually rub themselves on stones and branches or swim around frantically. Finally, they become lethargic, move to shallow water and die. The fungus growing on the lesions is a Phycomycete, *Saprolegnia*, and two species, *S. parasitica* and *S. ferax*, may be involved, the former being probably the most important. These fungi are also commonly seen to infect fish in domestic aquaria. Details of infection and the reasons for the sudden and catastrophic appearance of the disease in rivers are by no means fully understood. However, the spread of the disease is most marked during periods when the water temperature is low. It is also not clear whether *Saprolegnia* species are the sole causal agents or whether they act together with other, as yet unidentified, factors or organisms.

It is possible that fungal diseases of fish will become increasingly important because of the expansion of fish-rearing and fish-farming activities. Hatching and rearing tanks or lagoons, with their dense fish populations, provide good conditions for the epidemic spread of fungal pathogens. Reports of sudden and high mortality in both fry and adult fish are becoming more common and evidence is accumulating that hitherto rare infections are assuming a serious role. One of the most interesting of these is a fungal disease of the brain which has so far been only found in trout. The first apparent symptom is an erratic swimming pattern which is followed by lethargy and death. Mortality rates may reach 40 percent and outbreaks of this disease have recently occurred in commercial hatcheries in Canada. Such diseases provide new areas for research and the problems involved in their control are likely to formidable.

As with many other aspects of fungal behaviour, the deleterious effects of parasitic fungi on plants and animals can, in certain situations, be directed into channels which can benefit man. Alliances are possible in which parasitic fungi are turned against other pests of man, and fungi may even be induced to destroy one another. How this can be achieved is considered in the next chapter.

6

Allies against pests

There is something about killing rats which gives a deal of satisfaction ... almost any means of exterminating them is justified.
E. Samuel and J. Ivester Lloyd, *Rabbiting and Ferretting*, 1970

Uncontrolled agricultural pests can spell economic catastrophe and a large and growing array of pesticides is used against them. Pesticides are essential to maintain or increase crop yields and dependence on these chemicals is almost absolute. During recent years, biologists have begun to question the general acceptance of the situation that if a pest threatens a crop then a chemical must be used to control it. There are many reasons for this change in attitude, but a major factor has been the recognition that many agricultural chemicals are persistent. This means that they are broken down into harmless compounds only very slowly, or in some cases the breakdown compounds are themselves toxic, sometimes much more toxic than the parent chemical. Persistence has a high pesticidal value, since it increases the period over which killing is effective, but it also means that pesticides can gradually accumulate in a wide range of organisms, including man, until they eventually reach a harmful level. A second factor is the growing appreciation that in many situations a pest may be present in a crop but not at a sufficiently high level to cause economic loss. Nevertheless, the indiscriminate application of pesticides is commonly carried out in such situations, which is both economic and biological folly. Account is now being taken of the fact that all pests have either natural enemies, or may be susceptible to enemies artificially introduced into their habitat, and that these and other suppressive factors can be utilized in new systems of pest management. These systems embody integrated control in which chemical and biological methods of pest reduction are used in a compatible manner. The eradication of a pest is not essential as long as populations and their harmful effects are depressed below an economically important level. There are many examples of the spectacular use of biological control of pests—principally the destruction of weeds and of harmful insects—through the activity of natural or introduced enemies, these usually being plant-eating or insect-eating insects. Perhaps surprisingly, fungi too have proved to be effective organisms for biological control and their potential for this useful role is currently being actively studied. Fungi could be of value for controlling harmful insects, plant-parasitic fungi and pest weeds.

A great number of insects attack economically important plants but many insects in their turn are attacked by fungi. Some of these fungi may do little harm to the insect but others—entomogenous species—can cause severe or fatal mycoses. These fungi, with few exceptions, belong to one of two major groups; either to the Entomophthorales, which comprise an order in the Phycomycetes, or to the Fungi Imperfecti. Species in the Entomophthorales do not exist in a free-living form in nature but are always associated with their insect hosts. In contrast, species of Fungi Imperfecti can survive in nature as saprotrophs and some may live successfully in the absence of host animals.

Larvae as well as adults act as hosts for entomogenous fungi and infection is almost invariably effected by spores, the most common route of entry being directly through the host's cuticle. For spore germination to take place on the animal a high relative humidity, usually in excess of 80 percent, is required. Germ tubes arising from the spores grow for a short distance, then their tips swell and become firmly fixed to the insect's surface by means of a film of adhesive mucilage secreted by the fungus. Despite the hard and waxy nature of the exoskeleton of many insects, particularly adults, the fungus is normally able to penetrate successfully. The anchored germ tube produces a penetration hypha which is so fine that immense mechanical forces can be applied by its tip as it grows down towards the interior of the insect's body. The fungus literally drills its way through the protective covering of the host. The penetration tube may at the same time produce enzymes which can attack chitin and other components of the cuticle so that its physical integrity is destroyed and the process of penetration is made easier. Once the fungus reaches the soft, interior tissues of the body then growth accelerates. First, yeast-like cells which arise from fragmentation of penetrating hyphae multiply in the blood and become disseminated throughout the body tissues. These yeast-like cells then give rise to hyphae which penetrate the muscles and vital organs, destroy them and kill the insect. After death its body becomes rapidly filled with mycelium and, if conditions are favourable, hyphae emerge to the exterior and spores are produced upon them (Figs. 6.1 and 6.2).

The population size of any species of insect at any particular time depends on the resultant of two opposing groups of factors. The 'biotic potential' of the population, i.e. its capacity to increase in size, is continually modified by 'environmental resistance', which is made up of all those factors that act to reduce the size of the population. The biotic potential of an insect population which is inhabiting a crop is related principally to the initial size of the population, the proportion of fertile individuals within the population and their capacity to produce eggs. Environmental resistance contains both physical and biological components. For example, environmental resistance will be increased if climatic conditions slow down the development of eggs or reduce the movement or feeding activity of insects within the crop. Similarly, if natural enemies of the insect are present and conditions allow an increase in their numbers, environmental resistance will become more severe. The basis for studies of biological control methods that utilize insect-destroying fungi is the belief that such fungi are an important component of environmental resistance.

This belief is based on sound evidence, since field surveys carried out in various parts of the world have shown that populations of many economically harmful insects are affected to a significant degree by natural fungal infections. For instance, in the United States during the summer months 35–95 percent of looper caterpillars infesting soya bean crops may have fungi attacking them, while in the United Kingdom over 60 percent of female wheat-bulb flies may be killed by fungi before they are able to lay eggs. In the latter example it has been calculated that if insect-destroying fungi were absent, bulb-fly populations within a crop might in some cases rise to double the size normally observed. Despite the frequently high incidence of fungal infections which occur naturally in the field, in the vast majority of cases insect numbers still reach levels that require the intervention of chemical control. This is because, although epidemic spread of fungi throughout the population takes place, a sufficient number of insects escape infection so that biotic potential remains destructively high. In addition, natural outbreaks of disease often occur so late in the growing season that serious crop damage has already taken place well before the fungi begin to reduce population levels. Biological control, or the biological component of

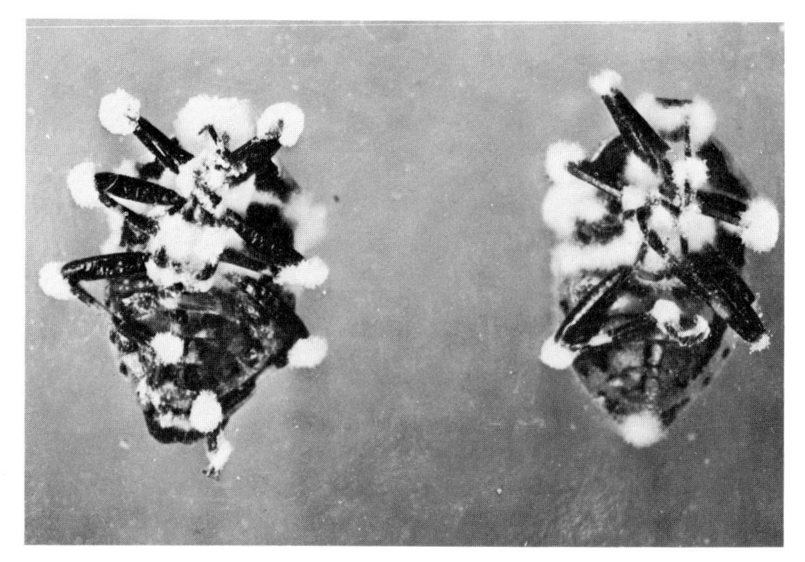

Fig. 6.1 Individuals of the two-spotted stinkbug killed by the entomogenous fungus *Beauveria bassiana* which has emerged through the intersegmental joints. (*Photograph by Dr E. Müller-Kögler, Pilzkrankheiten bei Insekten.* Paul Parey, Berlin and Hamburg, 1965.)

integrated control, therefore requires the development of a reliable method both for increasing the intensity of fungal infection within an insect population and for allowing epidemics to be initiated earlier than they might normally be. In order to achieve these ends large quantities of infective fungal material must be produced artificially and applied either in the form of spores or within infected, but still mobile, insect hosts. Raising and releasing immense numbers of insects at a suitable stage of infection is both expensive and time-consuming, and although a number of successes on a limited scale have been claimed this method of control is unlikely to be biologically or economically feasible in a wider context. The large-scale production and application of spores seems to be the most promising approach and for this procedure some species of the Fungi Imperfecti appear to be suitable as potential control agents.

Beauveria bassiana and *Metarrhizium anisopliae* are two such species that are easily

Fig. 6.2 Dorsal and ventral views of a fly killed by the entomogenous fungus *Entomophthora*. The pathogen has emerged to sporulate through the intersegmental joints. (*Photograph from Dr D. M. Macleod, Dr E. Müller-Kögler and Dr N. Wilding, Mycologia*, Vol. 68. 1976.)

grown in pure culture where, given suitable conditions, they will sporulate abundantly (Fig. 6.3). For more than 70 years numerous attempts have been made to control insect pests through the use of these fungi and results have ranged from impressive success to total failure. This unfortunate variability is due mainly to an almost complete lack of knowledge concerning the behaviour of the fungi under natural conditions. As well as this it is inevitable that once the fungus has been applied it may then become subject to adverse changes in environmental conditions that are beyond the experimenter's control. Lack of success might be due to too low a level of application; too low a population density of the host insect; the operation of temperature and humidity factors that reduce spore germination; a combination of two or more of these factors; or the effects of other, as yet unknown, factors.

Widespread and consistent success in the biological control of insects is currently being achieved by the use of applications containing bacteria, so that there

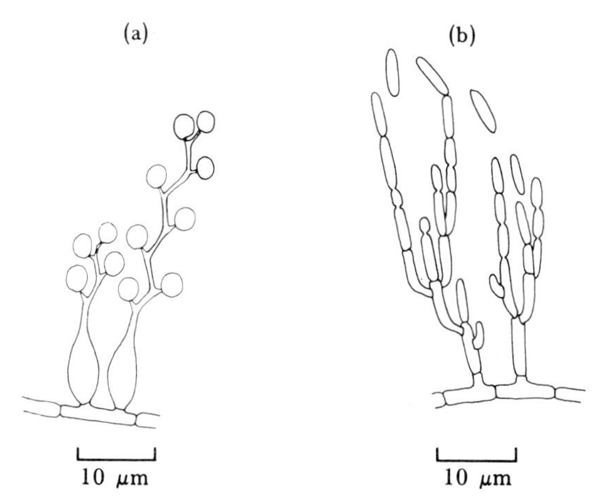

Fig. 6.3 Entomogenous fungi: (a) *Beauveria bassiana*, condiophores and conidia; (b) *Metarrhizium anisopliae*, conidiophores and conidial chains.

is some reason to suppose that eventually, after further fundamental research, fungi might be equally effective either alone or in combination with low levels of chemical insecticides. With *B. bassiana* and *M. anisopliae* there are no problems to be solved with respect to large-scale spore production since they grow rapidly and sporulate abundantly on cheap substrates which are easily handled and readily available, such as bran, molasses and starch. Spore yields range from 100–200 g per square metre of substrate, and this is not unreasonably low in relation to the amount of spore material required for agricultural application, usually 1–3 kg per hectare. Spores may either be dried and stored as a powder or frozen and stored without impairing their viability to any great extent. These advantages, together with the existence of technological resources for the continuous culture of fungi on an industrial scale, suggest that if biological control becomes feasible it need not be prohibitively expensive. Already commercial fungal preparations are available in the form of either wettable or dry powders for use in trials under field conditions.

The use of fungi as insecticides does, however, have its disadvantages when compared with the chemical control of insect pests. Some of these disadvantages, principally susceptibility to vagaries of weather, have already been mentioned. Other important shortcomings are lack of persistence and low speed of action. This makes repeated application necessary and increases the cost of crop protection. Chemical insecticides act rapidly, so that if high infestation is diagnosed or predicted then the pests are killed soon after application and the damage done by them is minimized. In contrast, development of a fungal epidemic is unavoidably much slower since the organism has first to infect the host, develop within it and then produce further infective spores. There is no way in which the time required for this process, commonly up to 3 weeks, can be reduced. In the meantime extensive crop damage might well have taken place. Even when entomogenous fungi have become active in an insect population their control value depends on the density of that population. Many fungi are most active when the host population is dense, and activity, as measured by the percentage of hosts infected, falls off steeply as the insect population falls. Fungi for biological control may therefore not be very efficient when pest populations are below a certain level, yet this level may still be one where economic crop losses might be caused. There is also the problem of specificity. Most entomogenous species of the Entomophthorales and Fungi Imperfecti are not host specific and some, particularly *B. bassiana*, have a very wide host range. This is, of course, of great biological value to the fungi in that their activities are not affected by the dearth of a particular species of insect. However, high specificity is a desirable characteristic for any pesticide since it allows the pest species to be killed without affecting valuable species or the natural enemies of the pest organism. Using non-specific fungi involves the same risks as using non-specific chemical insecticides. Finally, it should be noted that fungi may produce some unique and undesirable side-effects. *Beauveria bassiana* and *Metarrhizium anisopliae* have been found to be capable of causing lung infections in test animals and can induce severe allergic responses in some humans. There may therefore be some hitherto unforseen difficulties in the manufacture and application of fungal spore powders on a large scale.

Pests of cultivated crops are not the only insects which it might be desirable, if possible, to control biologically. Some species of mosquito are of great medical importance since they act as vectors of malaria and other serious diseases of man. Their immense breeding grounds, together with their concentration in predominantly underdeveloped parts of the world, require that any method of chemical control should be inexpensive. The well-known insecticide DDT has been utilized for this purpose for many years. The nature of the persistence of this chemical and its ability to accumulate in man and animals has received wide publicity and its use is now prohibited in some well-developed countries. However, it might be argued that it is difficult to justify an absolute ban on this and other persistent insecticides in the absence of other equally inexpensive control measures. In addition to problems of persistence there are other difficulties that arise from the regular and widespread use of DDT and similar chemicals. Populations of insects are emerging which are tolerant of very high levels of pesticides so that the effectiveness of chemical control is being reduced or completely negated in many areas. Interest is now being shown in the potential of fungi for mosquito control.

The stage in the life cycle of the mosquito at which biological control is directed is the larval phase. Larvae grow and develop in still water in large, concentrated and relatively sedentary populations. Favourable conditions for the epidemic spread of parasitic fungi would therefore seem to exist within these populations. Killing rates of 70–90 percent of larvae have been obtained by dusting the surface of breeding ponds with spores of *B. bassiana*. Even more effective and long-lasting results might be possible by using fungi that are specifically parasites of mosquito larvae and which can become established components of environmental resistance in breeding areas. One species has been discovered that might be effective in this way.

Coelomomyces is an aquatic fungus which attacks mosquito larvae and cannot apparently lead a free-living existence outside its body. Within the larva a thallus of wide hyphae is produced together with hyphal bodies that circulate in the insect's blood stream. As the fungus develops it exploits the host's fatty tissues and infected larvae normally die before they can pupate. Natural *Coelomomyces* infections normally occur at low levels or may even be entirely absent from some mosquito breeding grounds. It has been shown recently that the fungus can be successfully established in populations that are normally free from it and initial results indicate that such a method of biological control is not unpromising. It is, however, too early yet to be absolutely certain, since research is still in its preliminary phase.

It is curious that one of the most outstanding examples of the successful biological control of a serious pest by means of a fungus is not widely-known, even among professional biologists. Perhaps this is because the pest is itself a fungus and its effects, although they are certainly economically important, are not as obvious or dramatic as those produced by plagues of weeds or insects. *Fomes annosus* is a parasitic, basidiomycetous fungus that causes a variety of root-rot and butt-rot diseases of conifers. The fungus enters the small roots of healthy trees and proceeds to colonize their bark and wood, travelling within them until the larger roots are reached. In these it advances through the central woody tissues within which it causes extensive rotting. Eventually the base of the trunk is invaded, rotted and the tree dies. The parasite then forms its bracket-shaped fruit bodies on the dead or dying conifers and large numbers of basidiospores are released from these to be dispersed in air currents (Fig. 6.5).

Conifer plantations established on land which has been free of trees for a great number of years may remain unaffected by *F. annosus* until felling begins. Then, when young trees are felled, either during thinning or the cutting of access routes and firebreaks, *F. annosus* often appears. It is found first on the stumps of felled trees and later on the butts of standing trees. It has been shown that wind-blown basidiospores can infect the freshly-exposed wood of stumps and that the mycelium arising from the spores grows down into the roots. From there, infection of the healthy roots of neighbouring trees takes place by means of root-to-root contacts that act as bridges across which the mycelium can travel. Freshly-cut stumps allow the fungus to gain initial entry to an area, and from these bases it can then spread through stands of healthy trees. Obviously if the stumps could in some way be protected from basidiospores immediately after felling has taken place the risk of *F. annosus* infection would be eliminated. It is obvious also that the simplest method of doing this is to coat the stump with a compound that renders the surface layers of the wood unfavourable to fungal growth. Creosote has commonly been used to seal cut surfaces, but this treatment is effective for only a relatively short period of time. When the seal formed by the creosote-soaked tissues becomes cracked, as it inevitably does, then basidiospores are able to enter the underlying wood and infect it.

An important biological characteristic of *F. annosus* is that it is a specialized parasite which cannot compete successfully with other fungi when it encounters them in wood. The surface tissues of freshly-cut stumps are alive and can be colonized only by fungi, like *F. annosus*, which have the ability to overcome the residual resistance of the host and colonize living wood. In contrast, saprotrophic fungi can only colonize stumps when the tissues are moribund or dead. One method of control of *F. annosus* depends on encouraging the development of such saprotrophic fungi on cut surfaces so that when basidiospores of the parasite arrive they then either do not germinate or, if they do, the young mycelium arising from them is unable to compete with the established saprotrophic species and dies. Ammonium sulphamate applied to stumps kills the surface tissues and allows them to be entered by saprotrophs which exclude *F. annosus*. This is a good method of control that is partly chemical and partly biological. A more effective alternative is to inoculate the stump tissues with a fungus that is known to compete successfully with *F. annosus*. *Peniophora gigantea* is a parasitic Basidiomycete which, like *F. annosus*, invades freshly-exposed wood. It is, however, a weak parasite and is not a serious pathogen of forest trees. When spores of this species are sprayed onto stumps, the fungus rapidly colonizes them and prevents subsequent infection by *F. annosus* which is unable to compete with the established mycelium of *P. gigantea*.

There is another considerable advantage in using *P. gigantea* as a biological control organism. Where plantations within which *F. annosus* has been active are clear-felled and replanted with young trees, there is a high probability that *F. annosus* will attack the new trees when their roots come into contact with those of the old, infected stumps within which it can survive for considerable periods. This can be prevented by pulling out the old stumps and root systems mechanically, but this process is expensive and in some situations the terrain may be unsuitable for operation of the necessary stump-extracting machinery. In addition, roots which break off and remain in the soil can still act as sources of infection. If, when plantations containing infected trees are clear-felled, the stumps are inoculated with *P. gigantea* spores, then the fungus grows down into them. *Fomes annosus* is prevented from occupying and producing its fruit bodies upon the stumps. *Peniophora gigantea* can also advance through roots already colonized by *F. annosus* and eliminate the latter from them (Fig. 6.4). *Peniophora gigantea* can be grown in the laboratory where it produces abundant spores, and dehydrated tablets containing known numbers of viable spores can be prepared and stored. When required these tablets can be suspended in water and used in a spray. In the United Kingdom this method of biological control has been adopted by the Forestry Commission and has been used over a total plantation area of 28,000 hectares.

Fomes annosus is not the only tree-destroying fungus that can be controlled by the use of other fungi. The basidiomycetous species *Armillaria mellea*, the 'honey fungus', causes serious root-rot in both hardwood and softwood trees. As is the case with *F. annosus*, its spores can infect the freshly-exposed wood of stumps after felling has been carried out, and colonization of the stump tissues takes place. Some years after infection fruit bodies of *A. mellea* appear on the stump (Fig. 6.6). The fungus is not, however, confined to the stump and grows out from it through the soil and along its dead roots by means of hyphae which are aggregated into dark strands called rhizomorphs. Root disease breaks out in trees growing round infected stumps when these rhizomorphs contact healthy root systems and enter them. During their growth, rhizomorphs draw all the

Fig. 6.4 Section through a root from a Corsican pine stump showing the replacement of *Fomes annosus* by *Peniophora gigantea* from the bark inwards. The tree was naturally infected by *F. annosus* and, after felling, the stump surface was inoculated with *P. gigantea* which then grew from the stump into the root. The root was collected 20 months later, cut and incubated. (*Photograph by Dr J. Rishbeth, from Annals of Applied Biology, 1963.*)

Fig. 6.5 Fruit bodies of *F. annosus*. (*Photograph by Mr. R. Endall, Royal Botanic Gardens, Edinburgh.*)

nutrients that they require from the parent stump and are entirely dependent on it for persistent spread. In hardwood infections, the very persistence of the stumps allows *A. mellea* adequate time in which to extend these organs through the soil in search of new hosts.

For some time it has been forestry practice to treat freshly-cut hardwood stumps with ammonium sulphamate in order to control re-growth of shoots from them. This treatment, quite incidentally, also allows a system of natural biological control to act against *A. mellea*. As ammonium sulphamate penetrates the woody tissues it kills the living cells of the wood and also enriches the tissues with nitrogen. This action favours rapid and extensive colonization of the stump

Fig. 6.6 Hardwood stump bearing fruit bodies of the Honey fungus, *Armillaria mellea*. (*Photograph by G. Woods.*)

surface by saprotrophic, wood-rotting fungi, some of which can successfully compete with *A. mellea* and so exclude it. In future this kind of biological control might be reinforced or made more certain, where economically feasible, by also inoculating the cut surfaces with suitably competitive fungi. Alternatively, chemical treatments that selectively favour particular competing fungi may be developed.

With growing world demands for timber it is likely that problems caused by the activity of those stump-inhabiting fungi whose development is favoured by extensive felling will increase. Effective biological control of such fungi might, like that of *F. annosus* and *A. mellea*, provide an answer.

Weeds can cause enormous crop losses by physically choking out crop plants or competing with them for light and available nutrients and water. They can also harbour harmful insects or act as reservoirs for plant-parasitic fungi and nematode worms. An essential routine in farming is weed control, and use is made of the large range of selective herbicides now available to kill the weeds but leave the crop undamaged. In some circumstances, however, the use of herbicides may be both impractical and impracticable. A good example of such a situation is that involving 'skeleton weed' in south-east Australia. Skeleton weed, *Chondrilla juncea*, is a large, robust biennial plant that grows up to a metre tall and belongs to the Compositae, the same family as the dandelion and ragwort (Fig. 6.7). It is a native of the Mediterranean, Europe and North Africa and it was accidentally introduced into south-east Australia at some time before the

Fig. 6.7 *Chondrilla juncea*, the skeleton weed.

First World War. The plant spread rapidly and has become an extremely serious weed over vast areas of New South Wales and eastern South Australia (Fig. 6.8). The great areas involved, together with the high reproductive potential of the weed, make control through herbicides difficult and eradication of the plant by the same means impossible. In Europe *C. juncea* is attacked by a rust fungus, *Puccinia chondrillina*, that either kills the plant or debilitates it to such a degree that its growth and seed production are drastically reduced. *Puccinia chondrillina* only produces disease in plants of the genus *Chondrilla*, and in June 1971 it was imported into Australia in an attempt to control the activity and spread of skeleton weed. Plants which had been infected in the laboratory were transplanted into an area of dense skeleton weed at Wagga, New South Wales. The rust spread through the weed population by means of its airborne spores and by March 1972 had become widely distributed within the weed-affected areas. While the disease may not finally eradicate the weed, it has already been of immense benefit in reducing the infestation of cultivated land.

In addition to land plants, water-borne plants can also become noxious weeds, and in many parts of the world are providing an increasing threat to water resources and are creating ecological hazards on a large scale. Problems associated with aquatic weeds most commonly arise in the tropics and subtropics where profuse growth is favoured by the warm climate and long day. Newly-created lakes and canals are particularly prone to invasion, the most dramatic example in recent times being the rapid covering with the water lettuce, *Pistia stratioites*, of the lake formed by the Volta dam project in Ghana.

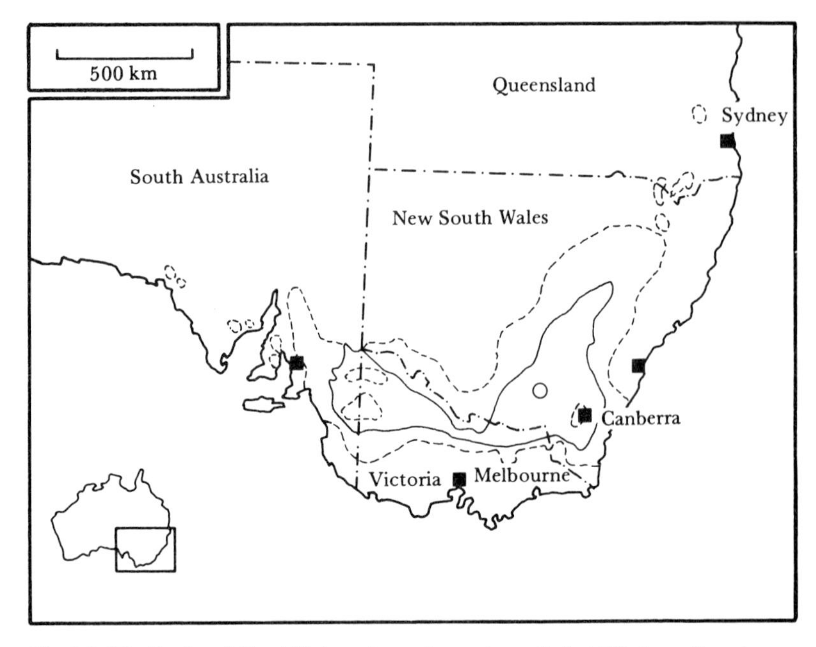

Fig. 6.8 Distribution of *Chondrilla juncea* in south-east Australia in 1972. Area of continuous infestation is delimited by the solid line, that of scattered infestation by the dotted line. The open circle indicates the point of release of *Puccinia chondrillina* at Wagga, New South Wales. (After Cullen, Kable and Catt, 1973.)

The environmental consequences of aquatic weed infestation range from the minor and irritating to the major and calamitous. Weeds quite obviously may clog the grids and sluices of irrigation systems and hydroelectric generating plants, they may make the navigation of inland waterways difficult and reduce all kinds of recreational activities. More seriously, they severely deplete water oxygen levels when they become senescent and decompose. They also—and this is not widely recognized—vastly increase water loss from reservoirs and irrigation canals by means of evapo-transpiration. For instance, a conservative estimate of the annual cost of increased water loss due to weed infestation in 17 western States in the United States has been found to be in the region of 39 million dollars.

Aquatic plants may also affect agriculture directly, particularly in zones where subaquatic crops such as lowland rice are grown, since they compete directly with such crops and also overrun land left to fallow. Finally, these weeds can threaten the health of human populations living close to infested water by providing refuges for snails and mosquitoes which are either vectors or alternate hosts of organisms causing malaria, filariasis and other diseases. Two obvious control methods that are used against aquatic weeds are the application of herbicides or, in restricted areas, mechanical removal of the plants. Both require to be carried out frequently, both are expensive and the continued pollution of water resources with herbicidal chemicals may in the long term create its own ecological problems. Biological control by means of herbivorous animals, plant-pathogenic viruses and bacteria, or parasitic fungi may in the future provide an alternative method. The search for suitable fungi has been in progress for a number of years and it is likely that it will soon be possible to control biologically one weed, *Eichhornia crassipes*, the water hyacinth, if only on a limited scale.

The water hyacinth is a free-floating plant that has raft-like foliage and hanging, submerged roots. It can become uniformly established over stretches of water and can adjust to fluctuations in water level. Although it produces flowers it does not set seed but is propagated asexually by the formation of offshoots. Reproduction is potentially very rapid, and it has been calculated that 10 individuals are capable of giving rise to 655,360 progeny in an 8-month growing season. Its great propensity for spread was demonstrated in 1952 when it appeared in the Congo River and subsequently colonized the 1,600 km stretch from Leopoldville to Stanleyville in a period of 3 years. Several fungi which attack the water hyacinth have been discovered, but with most the damage caused is slight or the progress of the disease is so slow that no control is effected. However, recent work in Florida gives cause for optimism.

The water hyacinth is a native of Central America and is an attractive ornamental plant with lavender flowers and exotic foliage. In New Orleans in 1884 plants were distributed to visitors to the New Orleans Cotton Centennial Exposition. These were subsequently grown in ornamental ponds and in a number of ways must have found their way to natural waterways. The plant reached Florida by 1890, and by the turn of the century was distributed as far west as California and as far north as Virginia. In Florida alone, 120,000 hectares of waterways are infested and the annual control programme costs over 15 million dollars. In 1971 a natural decline in the water hyacinth population was noted in the Rodman reservoir, a body of water along the Cross–Florida barge canal. Plants were found to be chlorotic, had rotting roots and failed to produce offshoots. All these symptoms increased in severity during the growing season and were found to be due to infection by *Cercospora rodmanii*, a species of the Fungi Imperfecti. This natural outbreak of disease declined in severity after 1971 until, by 1974, few diseased plants could be found in the population. There were, nevertheless, good indications that if *C. rodmanii* could be well established it might, either alone or in conjunction with herbicidal applications, provide a means of control. *Cercospora rodmanii* can be grown in the laboratory on artificial media, and limited field trials using its spores and fragmented mycelium in a spray were successful. Although a great deal of preliminary investigation remains to be done, there are indications of potential success.

It may be possible to use fungi against a number of important weeds, but the introduction of a fungal disease into a geographical region from which it is naturally absent carries some risk, so that a series of precautions must be taken.

A great deal of preliminary investigation has to be carried out over a long period of time before the introduction can take place with a minimum of danger. First, the propensity of the fungus to attack species other than the target plant has to be determined, in particular any cultivated plants that are closely related to the weed. Having done this, the process must then be repeated with the same varieties of plants but this time taking them from populations that have always been geographically isolated from the fungus. These may not have evolved resistance to it. In this way it is possible to avoid releasing a fungus that might transfer its activities from the target population to ecologically important non-crop plants or to valuable crop plants.

There is no doubt that the biological control of pests, albeit limited in scale, by means of fungi has an important future. By using these organisms in a controlled and beneficial manner, man might be said to be choosing his own symbiotic fungal partners. However, man is not the only animal to have done this and a number of invertebrate animals unmistakably owe their ecological success to their close association with fungi.

7

Secret gardens and insect hosts

There is a good possibility that in the twilight of this world the forms of life predominating will be the fungi and those plants and animals smart enough to have gone into partnership with them.
C. M. Christensen, *The Molds and Man*, 1965

Soon after Europeans arrived in the New World and began to establish their plantations they noted the serious damage that could be caused to crops by the activities of ants (Fig. 7.1). These dwelt in large nests into which they continually carried vast amounts of leaf material that they had cut from plants. Even now in Central and South America such ants are classified as plague animals and constant attempts are made to control their numbers. The ants belong to a single tribe, the Attini, and are restricted to South America east of the Andes, the Caribbean Islands, Central America and Mexico, and southern and eastern parts of the United States. The most important and conspicuous genus is *Atta*, some of whose species are commonly called 'Parasol ants' because of their appearance when they are carrying large leaves (Fig. 7.2).

Attine ants do not eat the leaf material that they gather but use it as a substrate on which they grow fungi for food. Their nests are subterranean, with numerous chambers, some of which contain fungus gardens that are constantly supplied with fresh leaves. In the gardens, abundant fungal mycelium grows on a spongy mass of decomposing organic matter, and each garden also contains a brood of ant larvae. The workers that are responsible for the supply of organic matter to the gardens belong to distinct castes each with a different series of tasks (Fig. 7.3). The largest ants are those of the maxima caste and these leave the nest to cut leaves but also guard the brood within the garden chambers. Smaller workers, the media, cut leaves but in addition tend the gardens and feed the brood. They divide leaves and large leaf pieces into smaller fragments, continually licking them and pressing the margins with their mandibles so that the tissue is pulped. Each fragment is anointed with a droplet of fluid exuded from the anus and is then inserted into the garden. Tiny ants, belonging to the minima caste, are restricted entirely to the garden, where they tend the fungus and care for the eggs and smaller larvae. As the fungus grows on its substrate it produces short, aerial hyphae which terminate in swollen, cytoplasm-filled organs called bromatia or gonglydia (Fig. 7.4). It is these nutrient-rich cells that are eaten

Fig. 7.1 Trail made by *Atta* workers which have cut and removed grass to create paths. (*Photograph by Professor N. A. Weber, from Gardening Ants – The Attines, 1972.*)

by the ants and are fed to the larvae. The condition of the fungus is ceaselessly checked by the workers who feel it with their antennae, lick it and frequently fertilize it with anal droplets. When new leaf fragments are added to the garden these are often inoculated with a minute tuft of hyphae picked up by a worker. Those ant fungi that have so far been carefully studied and identified have been found to be Basidiomycetes. They belong to *Agaricus* and other genera of mushrooms but they do not fruit within the nests and cannot apparently exist naturally in a free-living state outside them, although they can be grown in the laboratory. Only one species of fungus grows in the gardens of any given ant species.

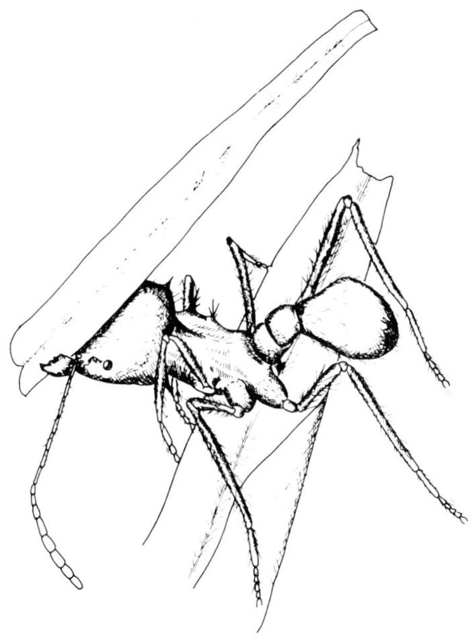

Fig. 7.2 An Attine worker carrying a piece of leaf material in the 'parasol' position. (After Weber, 1972.)

This maintenance by the ants of a pure culture of their food fungus is a remarkable achievement. Conditions within the garden are obviously non-sterile, and the presence of organic matter provides a suitable substrate for the growth of a large number of alien fungi. The spores of such fungi must continually be brought into the nest on the bodies of workers or on leaves so that the failure of contaminant moulds to grow requires explanation.

Unfortunately no generally acceptable explanation is available. The small size of minima workers may allow them to weed out foreign fungi but control or eradication by this method alone seems unlikely. Saliva and anal fluid may contain compounds which prevent the growth of contaminants without harming the food fungus, or the food fungus itself might also produce antifungal compounds. However, there is no experimental evidence that either of these explanations is correct.

The existence of only one fungus in the nests of any particular ant implies that each ant species is dependent on a single fungus for food. If gardens are exchanged between ant species they are frequently rejected, but this could be because the ants detect and react adversely to the previous activity of an alien ant species within the exchanged material. If laboratory-grown pure cultures of food fungi are used, these are acceptable, at least in the short term, to ants that normally do not grow them. It is interesting to note that some species of ant will readily accept a number of free-living species of the Basidiomycete *Lepiota*.

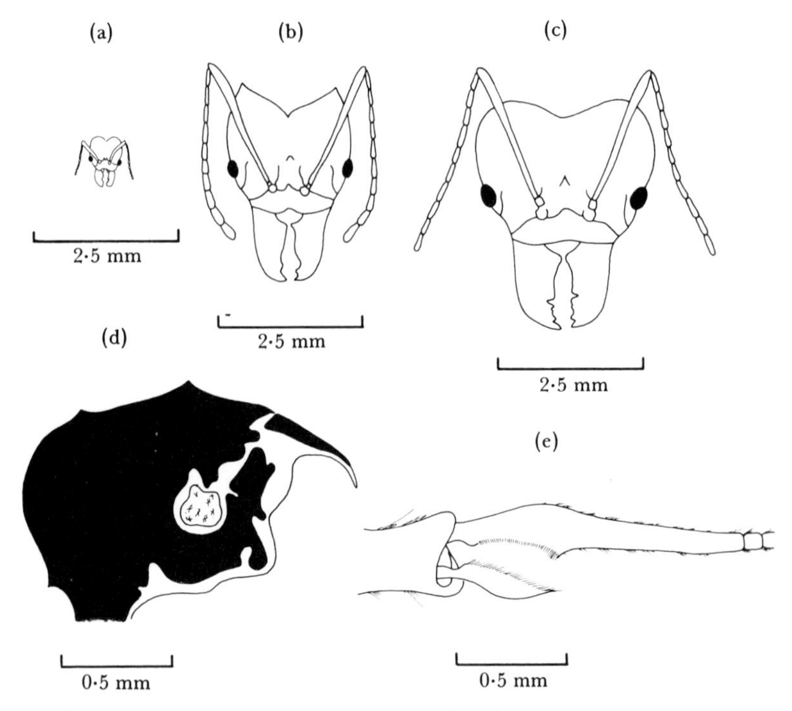

Fig. 7.3 Anatomy of Attine ants: (a)–(c) heads of minima, media and maxima workers, respectively; (d) vertical section through head of a queen showing the pouch containing a pellet of the food fungus; (e) part of the foreleg of a worker showing the 'comb' used for cleaning its appendages and grooming the fungus. (After Weber, 1966. Copyright American Association for the Advancement of Science, 1966.)

It has been shown that fungus-growing Attines ingest plant juices from the cut surfaces of leaves during both foraging and the preparation of the fungus gardens. However, it is not clear what proportion of the total food intake of the ants is provided by these juices. It is generally considered that the fungi provide an almost complete diet for Attine ants, who cannot survive without them. The ants depend ultimately on plant materials for all their nutrients but, unlike many other kinds of ants, they lack the ability to break these down and digest them themselves. The saprotrophic food fungi convert the indigestible substrate, principally cellulose, into proteins and carbohydrates, which the ants can then obtain by cropping the bromatia. In addition, some ant fungi have been found to be rich in a sterol, ergosterol, which the ant larvae may require for normal development. The fungus–ant relationship is a truly symbiotic one since in return the fungi, unable to survive outside the garden chambers, are given shelter and are provided with nutrients in the form of carbon-rich leaves. There is one curious additional feature in the symbiosis. In order to maintain luxuriant growth the food fungus requires a supply of nitrogen. This may be supplied directly to it in anal droplets, but the bulk of its nitrogen seems to be obtained from the leaves brought it to by the ants. Much of this leaf nitrogen is in the form of

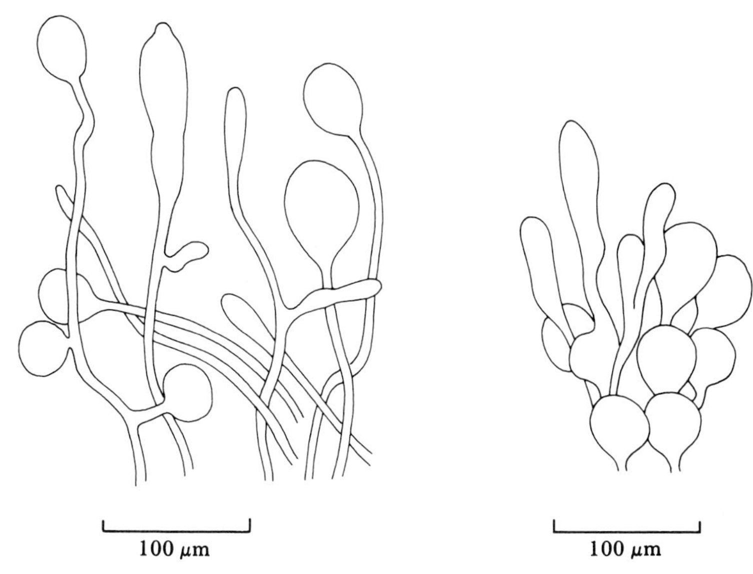

Fig. 7.4 Nutrient-rich bromatia of Attine food fungi. (After Weber, 1966. Copyright American Association for the Advancement of Science, 1966.)

proteins and polypeptides and, although it is capable of degrading complex carbohydrates, the fungus lacks the enzymes necessary to break down these nitrogen-rich compounds and so release simpler nitrogen compounds which it can readily assimilate. This enzyme aeficiency is made good by the ants, whose anal droplets contain the necessary enzymes (proteases) in abundance.

How the symbiotic, fungus-growing mode of life originated and developed is a complete mystery. Termites, a more ancient life form than ants, and one from which the latter probably evolved, also have fungi within their nests, but whether or not they feed on them to any great extent is a matter of debate, some considering that the fungi are a major nutrient source, others suggesting that they are not. If fungi are unimportant as a food, they might instead help to produce an elevated temperature within the termite nest (the termitarium) through their saprotrophic activities and perhaps also maintain a high relative humidity. It has been suggested, rather speculatively, that ants could have accidentally derived their fungus-growing habit from an encounter with termites.

Ants invading a poorly-defended or enfeebled termite colony might have come upon the saprotrophic fungi growing within its chambers, found them to be palatable and transferred them to their own nests. Then, during successive generations, the fungus-culturing way of life gradually became more and more important as reliance on other sources of organic foods was gradually lost, culminating in absolute dependence on the fungi for survival. Termite fungi, perhaps erroneously, have been referred to as the stomach and liver of the termitarium but this is certainly the function of the ant fungi within the Attine nest. In an alternative view of the evolution of the fungus-culturing habit it is proposed that the symbiosis arose quite independently. It is suggested that ancestors of the Attines deposited their waste at special sites in or near the nest and that fungi then developed on this. The nature of the waste material allowed only a few fungus species to grow upon it, and some of these were capable of digesting cellulose. Ants then discovered that these species could be eaten, and gradually became more and more nutritionally dependent on them. It was then one small, but significant, step to the habit of growing the food fungi by intent and developing a method of transferring them to new nest sites.

Absolute dependence on the food fungus requires that when a new nest is built the fungus is carried to, and established within, the pioneer ant colony. Nests of *Atta* species are founded by fertile queens following their nuptual flight. Each queen has a small pouch, situated towards the back of the mouth, which contains a small pellet of mycelium of the food fungus (Fig. 7.2). The ant lands on the ground and sheds her wings and excavates a small chamber. She then occupies this and immediately begins to lay eggs. These are of two kinds—'alimentary eggs', which she consumes to maintain herself, and 'reproductive eggs' which are capable of producing larvae. The queen plants the fungus and feeds it with anal droplets so that a primordial garden develops. She does not, however, eat the fungus at this stage and several months elapse between the laying of the first reproductive eggs and the emergence of the first workers, during which the queen eats only her alimentary eggs. The workers begin to tend the garden, gather leaves for its expansion and feed the queen on bromatia (Fig. 7.5). The laying of alimentary eggs then ceases.

Leaf-cutting ants are economically harmful because of their destruction of crop plants. However, in non-agricultural areas they are ecologically beneficial due to enrichment of the soil by the enormous amounts of organic matter which the workers carry into their subterranean nests during the several years they are occupied. For example, within a mature nest there may be over 1,000 chambers and during their excavation as much as 40,000 kg of soil can be carried to the surface. About a third of the chambers contain gardens and the soil removed to create these is replaced by fresh organic matter. During the life span of a colony over 12 parts of weight of leaf tissue may be carried into the garden chambers for every part of soil removed from them. Over a period of 6–7 years this organic matter can amount to nearly 6,000 kg, and ultimately enriches the soil below the ground surface. In tropical rain forests few animals or tree roots normally penetrate to any distance below the surface, so that the activities of the Attines create important organic matter-rich habitats which otherwise would not exist. Their total impact on soil nutrition can thus be considerable.

Other insects also depend on plant materials for their nutrients and, like the Attines, lack the ability to break them down and digest them. These too use food fungi to overcome these shortcomings but they do not instinctively cultivate them or care for them. They have, however, evolved mechanisms which ensure that they are always accompanied by the fungi on which they depend for survival. Many of these insects inhabit wood which, either living or dead, provides a favourable habitat for insects that are able to bore into it and so create an abode. However, wood is an unpromising source of food since its main components are cellulose and lignin, and insects do not normally possess the enzymes necessary to degrade these materials. Wood is also deficient in vitamins and sterols that are essential for the normal development of insects but cannot be manufactured by them. The success of many wood-inhabiting insects is attributable to symbioses with fungi that can meet all their nutrient requirements and one of the

Fig. 7.5 The queen of an *Atta* colony and her workers attending a young fungus garden. (*Photograph by Professor N. A. Weber, from Gardening Ants—The Attines, 1972.*)

most successful groups of such insects is the wood-inhabiting Scolytid beetles (Fig. 7.6).

Fungus-eating Scolytids attack low-vitality wood such as recently-fallen timber. Flying pairs of adults alight on its surface and the female begins to bore into it. After penetrating for a short distance the beetle ceases boring, emerges from the tunnel and copulates with the male. Boring is then resumed by the female which excavates deep breeding tunnels or chambers and lays her eggs within them (Fig. 7.6). The female possesses special organs called mycetangia, consisting of pouch-like invaginations of the body surface within which either fungus spores or yeast-like cells are contained (Fig. 7.7). The cells are suspended in an oily liquid and are retained within the mycetangium by means of spines or hairs that protrude from its walls. The mycetangial liquid is secreted by glandular cells and may act as a source of nutrients for the fungi contained within it, while at the same time inhibiting the growth of other microorganisms which might gain access to the mycetangium. Secretion of the fluid increases as the

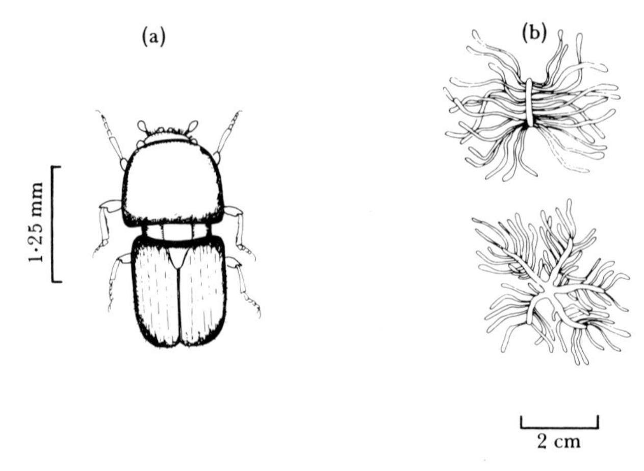

Fig. 7.6 Wood-boring beetles: (a) dorsal view of an adult Scolytid; (b) diagrams of beetle galleries within timber.

female tunnels, so that it oozes out of the mycetangia and deposits fungal spores or cells on the tunnel walls where they begin to grow. It is thought that the mycetangia originally had the primary function of providing a lubricant for the insect's body during boring, and perhaps also acted as a water repellent to protect the insect when it was passing through sap-rich wood. As evolution progressed, fungi that could tolerate or utilize the secretions began to occupy the tunnel walls and then invaded the secreting organs themselves.

After deposition of the fungus a mycelium develops which thickly lines the tunnels or chambers. The adult beetles constantly browse on this mycelium, at the same time keeping the tunnels free from debris and excrement. When the eggs hatch the larvae also eat the fungus until the pupal stage is reached. After

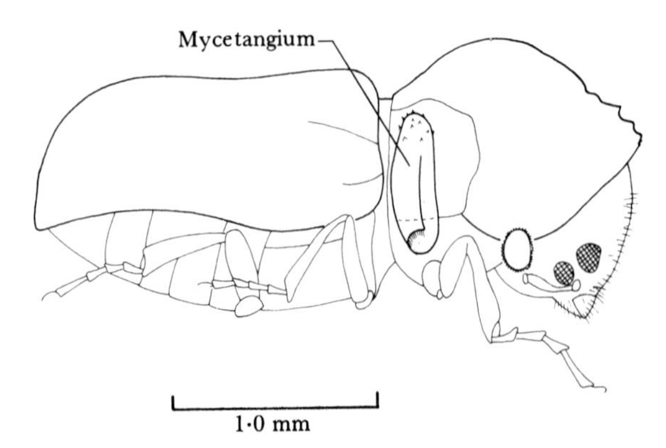

Fig. 7.7 Lateral view of a Scolytid beetle to show the position of the mycetangia. (After Francke-Grosmann, 1967. Copyright Academic Press Inc., 1967.)

pupation the young adult beetles feed for a time on the fungus then pass out of the wood, fly to a new site, excavate new tunnels and mate. Until a mycelium is established within the new tunnels the beetles do not feed but derive all their nourishment from absorbing their flight muscles. How the fungi enter the mycetangia of young adult beetles is not known.

The fungi involved in the partnership with these beetles are called 'ambrosia fungi' and are usually Ascomycetes or Fungi Imperfecti. Within the tunnels of any beetle species more than one species of ambrosia fungus can normally be found, and it seems that many beetles are not restricted to eating a specific ambrosia fungus or group of ambrosia fungi. Hyphae penetrate the tunnel walls and have a well-developed ability to break down wood to maintain their growth. Ambrosia fungi provide readily-assimilable major nutrients for adults and larvae, together with ergosterol which is necessary for larvae to develop into adults.

The benefits conferred on the Scolytid beetles from association with ambrosia fungi are clear, but in return advantages also accrue to the fungi. It is unlikely that most ambrosia fungi can exist in a free-living state outside the brood tunnels and mycetangia of their beetle partners. The animal thus ensures continuation of the life of these fungi and inoculates them directly into a substrate suitable for their rapid development. In addition, the beetle may provide them with some of the nutrients they require which may not be abundant in wood. There is evidence that some fungi may require fats, and these are present in the mycetangial secretions that are smeared on the tunnel walls. Fungi also have a high demand for nitrogen and in ambrosia species this might be wholly or partly met by the deposition of uric acid, a major excretory product of the beetles, on the mycelium.

A rather more loosely-defined association is that found between fungi and some species of Siricid wood wasps. Siricid larvae inhabit weakened trees or freshly cut timber and hatch from eggs inserted into the wood by the adult female with the aid of a long ovipositor (Fig. 7.8). The larvae then tunnel more deeply into the wood, growing and developing in debris-packed tunnels. The debris and the tunnel walls are permeated by fungal mycelium. Adult females possess, at the base of the ovipositor and opening into it, two pouches filled with mucus in which short, branched fungal cells are suspended (Fig. 7.9). The cells are retained within each pouch by means of inward-pointing hairs situated around the exit. When eggs are being laid the walls of the pouches contract so that fungal cells are squeezed out, contact and adhere to each egg, and become inserted into the wood with them. The cells then rapidly give rise to a mycelium which ramifies throughout the wood, destroying both its cellulose and lignin. The eggs hatch and the larvae become active in tissues which have been radically altered, both structurally and biochemically, by the action of the mycelium: the wood is softer and also probably contains higher than normal levels of short-chain carbohydrate molecules.

How the wood wasp larvae benefit from the presence or activity of the associated fungus is not clear. They do have the enzymes necessary to digest mycelium but the hyphae are so sparse in the wood that it seems unlikely that they could act as a major food source for the larvae. The most likely explanation is that the larvae are deficient in wood-degrading enzymes but are able to digest wood that has been somewhat modified by extracellular fungal enzymes. Action of the latter certainly makes it easier for larvae to move through the wood.

Whatever the exact relationship between the wood wasp and the fungus, its importance to the wasp can be inferred from the complicated mechanism which

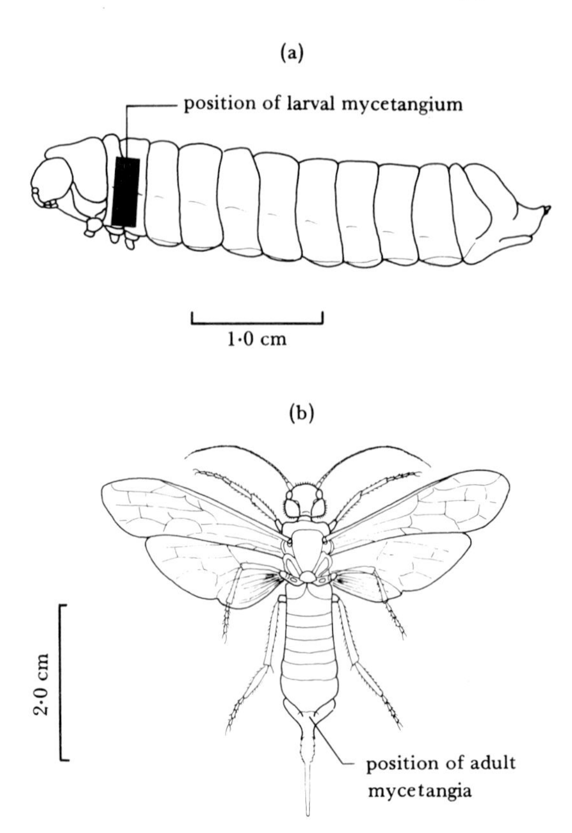

Fig. 7.8 Wood wasps: (a) lateral view of a Siricid larva, the black rectangle indicating the position between the body segments where the mycetangia are located; (b) dorsal view of an adult female showing the long ovipositor and position of the mycetangia.

has been evolved to ensure that the fungus finds its way into the ovipositor pouches of each new generation of adult females. Full-grown female larvae have a pair of mycetangia, one on each side of the body, hidden in a deep fold at the end of the first abdominal segment (Fig. 7.8). Each mycetangium consists of a series of deep pits full of an oily fluid within which fungal cells are suspended (Fig. 7.9). How the larval mycetangia are formed and how the fungus finds its way into them is not known. Shortly before the female larva begins its final moult, copious secretion of mycetangial fluid takes place and the secretion hardens to form small, waxy plates within which fungal cells are embedded. These plates are then shed. When the young adult female emerges from the pupa she begins to make continuous retractive movements of her ovipositor. This behaviour is instinctive, presumably being determined genetically, and draws fragments of plate material up into the already-formed fungal pouches at the base of the ovipositor. On reaching these the plate matrix dissolves, the fungal cells are released within the pouches and begin to grow.

Wood wasp fungi are easily isolated from egg surfaces or larval tunnels and

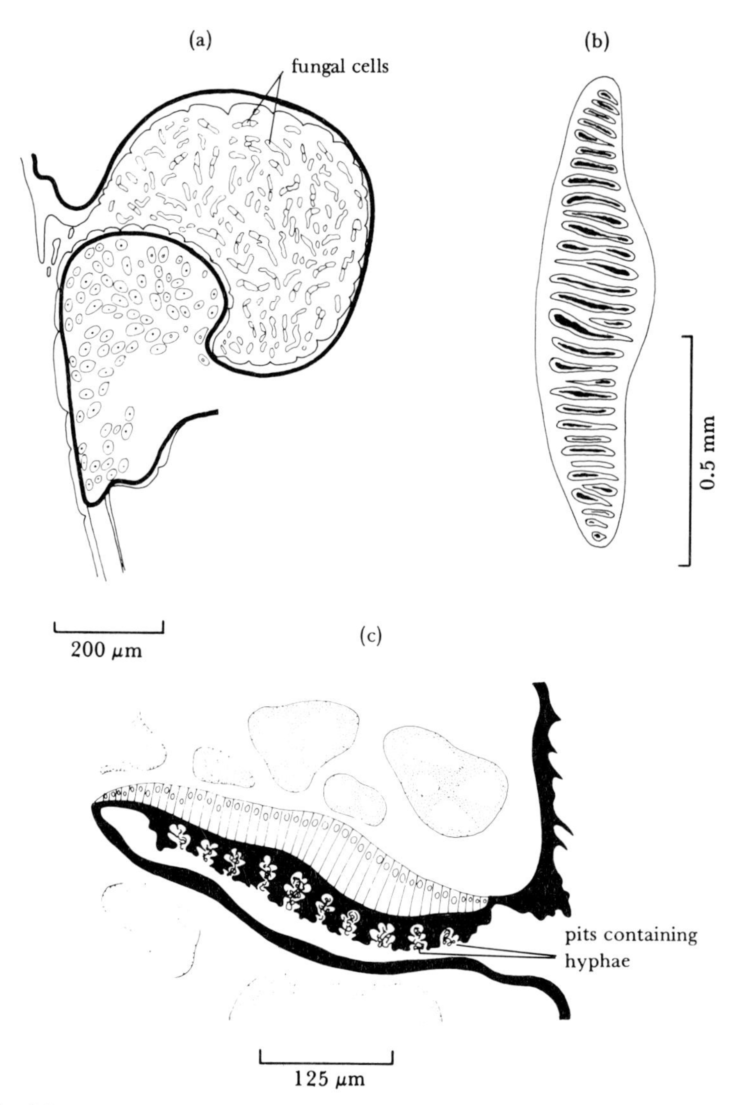

Fig. 7.9 Structure of mycetangia in wood wasps: (a) vertical section through the base of the ovipositor showing a fungus-filled pouch; (b) surface view of a larval mycetangium with slit-shaped pits; (c) section through a larval mycetangium showing bundles of hyphae retained within the pits by spines. (After Buchner, 1965.)

a number of them have been identified. They have usually been found to be species of the common and free-living Basidiomycetes *Amylostereum* or *Stereum*, which are wood-rotting fungi of some economic importance since they cause spoilage of commercially felled timber. The wasps must, in a great number of instances, act as vectors for these fungi, moving them from place to place and inoculating them deeply into wood. This appears to be the only advantage which the fungus obtains from the symbiotic association.

The symbiotic fungi of wood-inhabiting beetles and wasps are housed, and even grow, within the body of the host in special organs but they do not enter the cells of the body, always remaining outside them. However, in the most well developed fungus–insect associations the fungi enter and develop within host cells. Such fungi have so far been found mainly among leaf-hoppers, aphids and some kinds of beetle. The symbioses have great diversity but for the greater part they are at the moment poorly understood. The most comprehensive studies have

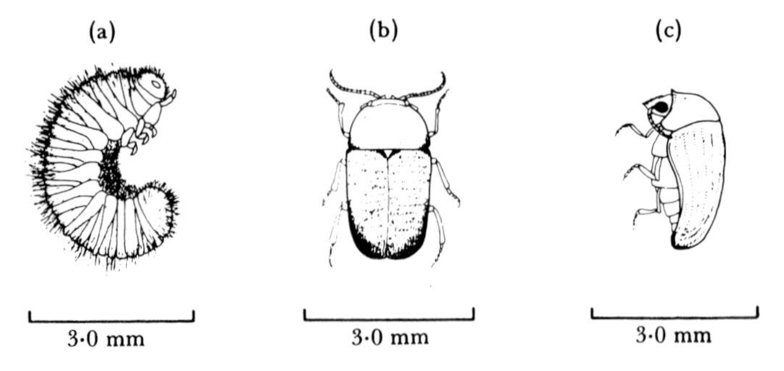

Fig. 7.10 *Sitodrepa panicea*, the drugstore beetle: (a) larva; (b)–(c) dorsal and lateral views of an adult.

been carried out on a single beetle species, *Sitodrepa panicea*, the 'drug store beetle' (Fig. 7.10). This insect, which is only 2–3 mm long, is an important food-spoilage organism, particularly in carbohydrate-rich products such as flour and rice. *Sitodrepa panicea* larvae have in their digestive system large mycetomes located in the region where the foregut and hindgut meet (Fig. 7.11). These mycetomes are modified parts of the gut and remain connected to the alimentary canal. The tissue which lines them is composed largely of special swollen cells, mycetocytes, the cytoplasm of which is packed with actively budding yeast cells (Fig. 7.11). Adult beetles also possess gut mycetomes but these are much smaller than those of the larvae. Continuity of the symbiosis from adult to larva is ensured by a transmission organ mechanism morphologically similar to that found in wood wasps. The adult female possesses tubules which are continuous with the ovipositor and contain yeast cells suspended in fluid. When the eggs are laid they become contaminated with the fungus. In addition, at the end of the ovipositor are two more yeast-filled pouches which also inoculate the eggs with the fungus. The first activity of larvae when they hatch is to eat the egg cases from which they have emerged. Yeast cells are in this way ingested and pass undigested through the gut until they reach the region between the foregut and midgut. Here the mycetomes, which are already partly formed, await their arrival and fungal cells pass into the mycetocytes and begin to bud.

One problem remains, namely the infection in the new generation of adults of the organs which supply the fungus to the ovipositor and so smear the eggs. When the larva becomes transformed to the pupa these organs have been formed but are empty. After the young adult has emerged from the pupa the ovipositor tubules and pouches become inoculated with yeast cells at its first defecation. Since the hindgut opens into the ovipositor sheath the fungal cells can reach the transmission organs during this activity.

Within any group of insects the possession of symbiotic yeasts by particular species is normally related to their feeding habits. Species which consume some kinds of dead organic matter, or suck plant sap, or have a high solid carbohydrate intake, subsist on diets that are generally low in nitrogen compounds and may

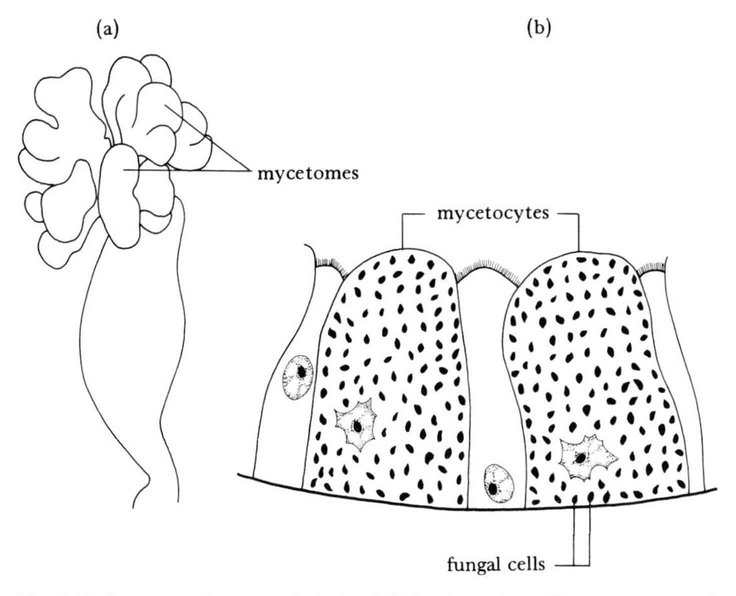

Fig. 7.11 Structure of mycetangia in Anobiid beetles: (a) sac-like mycetomes at the junction of the foregut and midgut; (b) section through the cells lining a mycetome showing the enlarged mycetocytes containing fungal cells. (After Buchner, 1965.)

also be vitamin deficient. The fungi in their mycetomes can make good these deficiencies, allowing utilization of otherwise inadequate food sources and the consequent occupation by the insect of nutritionally marginal habitats. For example, it is possible to obtain fungus-free larvae of *Sitodrepa* by surface-sterilizing eggs and allowing them to hatch in aseptic conditions, but on unsupplemented carbohydrate-rich diets the larvae grow at a much slower rate than is normal and there is high mortality. They also have difficulty in moulting during their development so that adulthood is rarely reached. If fungus-free larvae are provided with sterols and B-group vitamins in their diet, they grow and develop just as they would if they carried their symbiotic yeast within them. This implies that larvae in the symbiotic state are supplied with these compounds by their fungus. In addition, fungus-free larvae are deficient in many amino acids, the essential building blocks for proteins, and their fungus partially alleviates this deficiency in symbiosis. The supply of amino acids is created by the fungus from the host's nitrogenous waste products, principally uric acid.

While it is easy to understand the advantages obtained from the symbiosis by the insect it is difficult to visualize with certainty the benefits that might accrue to the fungus. It is probable that mycetome yeasts cannot exist in a free-living state in nature and that the mycetocytes provide a specialized congenial environment for their development. The remarkable tolerance of host cells to occupation by yeasts and the resistance of the latter to the effects of the host's enzymes, to which they must be constantly exposed, emphasize the highly-developed nature of these kinds of symbiotic association. This in turn implies that, in evolutionary terms, the associations must be very ancient to enable such a degree of mutual tolerance to have become widely established. Essentially the host and its fungus comprise a remarkably successful dual organism analogous to a lichen, except that in this instance both partners are heterotrophs.

The most bizarre of all the known fungus–insect symbioses is the association between basidiomycetous fungi and some species of scale insect. This kind of association is additionally interesting since it also involves the participation of a higher plant, normally a shrub or tree, upon which both fungus and insect ultimately depend for their survival. Species of the Basidiomycete genus *Septobasidium* grow on branches, but development is superficial and they rarely penetrate the plant. They have perennial, flattened thalli which frequently resemble those of lichens. Within each thallus is a labyrinth of chambers and tunnels opening to the surface by means of numerous entrances. Scale insects move freely through the maze of tunnels, but those in the chambers, usually occupied by a single insect, are not free to move and lie in direct contact with the bark of the plant (Fig. 7.12). Each is held by hyphae that are attached to the roof and walls of the chambers and pass through the insect's outer tissues into its body cavity, or haemocoel, which is filled with blood. Within the blood the hyphae bear tight, terminal coils. Scale insects feed by sucking sap, and those within the chambers feed by inserting their sucking mouthparts through the bark into the sap-rich layers of the branch. The insect feeds, its blood is supplied with nutrients and these are continually removed from the blood through the coiled hyphae of the fungus. The nutrients are then moved through the hyphae exterior to the insect's body and are distributed to the thallus tissues. The insect is thus used as a bridge for the passage of nutrients from plant to fungus, the latter being unable to directly exploit the plant itself. The imprisoned insects cannot breed and do not grow to full adult size, but they do remain alive and may even achieve a significantly greater age than those that are not penetrated by the fungus.

Continued growth of an established thallus or the initiation of new thalli on clean bark depend on a supply of fresh insects. This is provided each spring by those individuals that are moving freely inside the thallus when the fertile females among them give birth to young scale insects. This vernal production of young coincides with the formation on the thallus surface of basidiospores. These germinate to produce sticky cells that bud profusely in a yeast-like fashion. Young insects either remain within the tunnels, where they reach adulthood and subsequently breed, or they may crawl to the thallus surface and come into contact with adhesive bud cells which then proceed to infect them. During the initial stages of infection these insects may either re-enter their parent thallus or crawl to uncolonized bark. Those individuals that re-enter become responsible for maintaining growth of that thallus, while those that migrate establish new thalli. In either case the infection process is the same. The attached bud cells produce hyphae which enter the insect's body by growing through one of the pores from which the insect's hairs emerge. Growth proceeds until the haemocoel is reached and then, within this, the hyphae form their absorptive coils. When these are formed hyphae also emerge from the insect. If the insect has entered a chamber in an established thallus, the external hyphae link up with hyphae that have, in the meantime, grown down towards the insect from the roof and walls. The insect is then exploited and helps to maintain growth of the thallus. If the insect has moved to a new habitat the external hyphae grow, branch and form a small thallus containing the insect in a chamber. Other infected insects may then settle

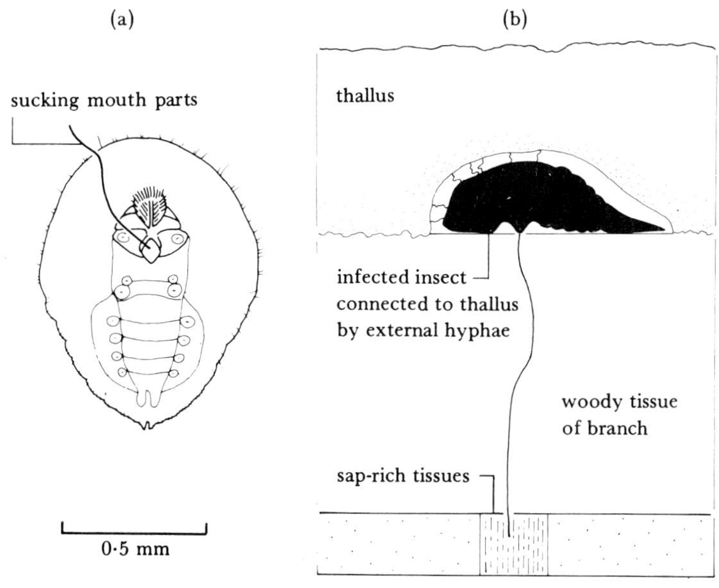

Fig. 7.12 Scale insects and *Septobasidium*: (a) ventral view of an adult insect showing the elongated, sucking mouthparts; (b) diagram of a section through a thallus of *Septobasidium* showing a single, infected insect within its chamber with its mouthparts inserted into the sap-rich tissues of the tree.

down in contact with this, and uninfected insects may shelter within it so that a new fungus-insect entity is initiated.

The *Septobasidium*–scale insect association results in a situation where there is mutual benefit to both partners, but it also contains factors which operate adversely on both partners. The scale insects can live unassociated with the fungus, although in nature they probably rarely do so. The thallus protects them from unfavourable environmental extremes and also gives them protection against their natural enemies, principally parasitic wasps. The price paid for this protection is the sacrifice of that proportion of the population which remains imprisoned in the fungal chambers. The fungus cannot survive in nature without the insects as it cannot obtain the nutrients that it requires directly from the plant. In addition, the scale insects provide its only means of dissemination, since its basidiospores cannot give rise to new thalli except through the agency of infected animals. However, infection need not always result in a successful fungus–insect partnership, since it has been observed that the host's white blood cells are sometimes capable of eliminating the fungus by destroying the coiled hyphae in the haemocoel. This complicated symbiosis allows each partner to occupy a habitat which neither could colonize independently, but the degree of dependence of the fungus on the insect is, perhaps, marginally greater than that of the insect on the fungus.

At certain times man, like some insects, practises mycophagy and consumes fungi as food. However, unlike insects, and possibly a great number of other animals, man is capable of error in selecting suitable species. It is upon past and present errors of this kind that the study of fungal toxins is mainly based.

8 Toxins

To conclude, few of them are good to be eaten and most of them do suffocate and strangle the eater.
John Gerard, *The Herball*, 1597

The collection and culinary use of wild fungi has its roots in antiquity. Their various flavours and textures have long been a delight to the gourmet, and their abundance at different times of the year has in the past provided a valuable supplement to an otherwise monotonous diet. An obvious problem with eating wild fungi is that there is a risk of poisoning through gathering non-edible fruit bodies by mistake. As a fifteenth century author pointed out: 'The Mushroom or Toadestoole ... hath two sundrie kinds, ... for the one may be eaten: the other is not to be eaten'. This division of fungi into the edible and non-edible is not strictly valid as the two kinds represent opposite extremes between which the majority of fungi lie. Most fungi may not be particularly pleasant to eat but they do not always produce adverse effects in the eater. In addition, there is no absolute definition of edibility. For example, some of those fungi that are prized as esculents may affect some people unpleasantly, particularly if taken in excess or with alcohol, and even the well-known mushroom of commerce, *Agaricus bisporus*, is toxic to laboratory animals if it is given to them as their sole source of protein.

In general, mushrooms and toadstools have a bad reputation with respect to their capacity to poison, but in fact fatalities are relatively rare. Available statistics are scattered, and may not be completely reliable, but for example, in the United States from 1965–7 there were 1,753 reported cases of suspected mushroom poisoning but no fatalities, and over the period from 1931–68 there were only 64 such deaths. In North America, fatal and non-fatal poisonings due to fungi probably account for less than one percent of the total annual cases of accidental ingestion of poisonous substances, although in the United Kingdom, Continental Europe and Japan this proportion seems to be much higher.

Mushroom poisoning, or mycetismus as it is known clinically, is a term that covers a wide spectrum of effects, ranging from a mild allergy or gastrointestinal responses to very severe illness, but the majority of cases are not serious. Severe symptoms are produced by relatively few fungi and among these species the two best-known are probably *Amanita phalloides*, the 'death cap', and *A. verna*, the 'destroying angel'. Unfortunately, there are no specific antidotes of proven therapeutic value as yet available for treating instances of poisoning by these and other *Amanita* species. After ingestion of *A. phalloides* the appearance of symptoms is delayed for 6–24 hours. There is then violent vomiting and diarrhoea, which lead to dehydration and a fall in both the blood pressure and the blood sugar level. If the victim survives during this primary phase there may be a remission, but this is frequently followed by relapse so that coma and death occur within 3–5 days. Because of the characteristic quiescent period between ingestion and the onset of symptoms, gastric or intestinal lavage applied to remove the meal on the appearance of symptoms is usually useless, since the poisons have already entered the blood stream by the time these measures are taken. Replacement of lost fluids, blood dialysis, or blood transfusions, are the major treatments available but none is guaranteed to succeed. However, although *A. phalloides* is widely considered to be the most deadly mushroom known, recoveries do take place, and mortality rates are between 34–63 percent.

Amanita phalloides, together with a number of other highly poisonous Basidiomycetes, contains compounds known as protoplasmic toxins that are capable of bringing about the degeneration of some of the major organs of the body. However, their role in clinical *Amanita* poisoning is not fully known and their nature and mode of action are still the subject of research and debate. At least 10 protoplasmic toxins can be extracted from the fruit bodies of *A. phalloides* and these can be divided into two roughly equal groups, phallotoxins and amatoxins. All these protoplasmic toxins are chemically related, being cyclopeptides made up of amino acids. The two most extensively studied compounds are phalloidin, a phallotoxin, and α-amanitin, an amatoxin (Fig. 8.1).

Phalloidin accumulates specifically in the liver and attacks the plasma membranes of the liver cells. The ability of these cells to control the movement of ions is badly impaired, so that there is at first a leakage of calcium from them, followed shortly afterwards by a massive efflux of potassium. Phalloidin may also affect other membranes within the cell, and the rupture of organelles may lead to the release of enzymes into the cytoplasm. The cell thus begins to digest itself and is destroyed. Affected cells become grossly distorted and their cytoplasm protrudes in areas where their plasma membranes have been weakened. Due to accumulation of blood within it the whole organ becomes enlarged and brittle and degenerates. Despite the drastic effects of phalloidin on liver cells, there are strong doubts as to whether this toxin is responsible for even some of the symptoms of *Amanita* poisoning. When introduced into the blood-stream of laboratory animals by means of intraperitoneal injection its effect is almost immediate. However, the poisoning of animals by orally administered phalloidin has never been demonstrated and it appears that the toxin is not efficiently transported from the stomach to the blood-stream.

Unlike phalloidin, α-amanitin is toxic when ingested. It causes lesions in the cells of the stomach, and in several regions of the intestine, and these probably give rise to the early gastrointestinal symptoms. This damage, however, is relatively minor in comparison with its subsequent effects when it reaches the liver and kidneys via the blood-stream. Within a few minutes of entering cells, their nuclei are attacked and the nucleoli begin to disintegrate. Nucleoli are structures located within cell nuclei and are composed largely of ribose nuclei acid (RNA). This material is essential for the cell to manufacture protein, so that when the

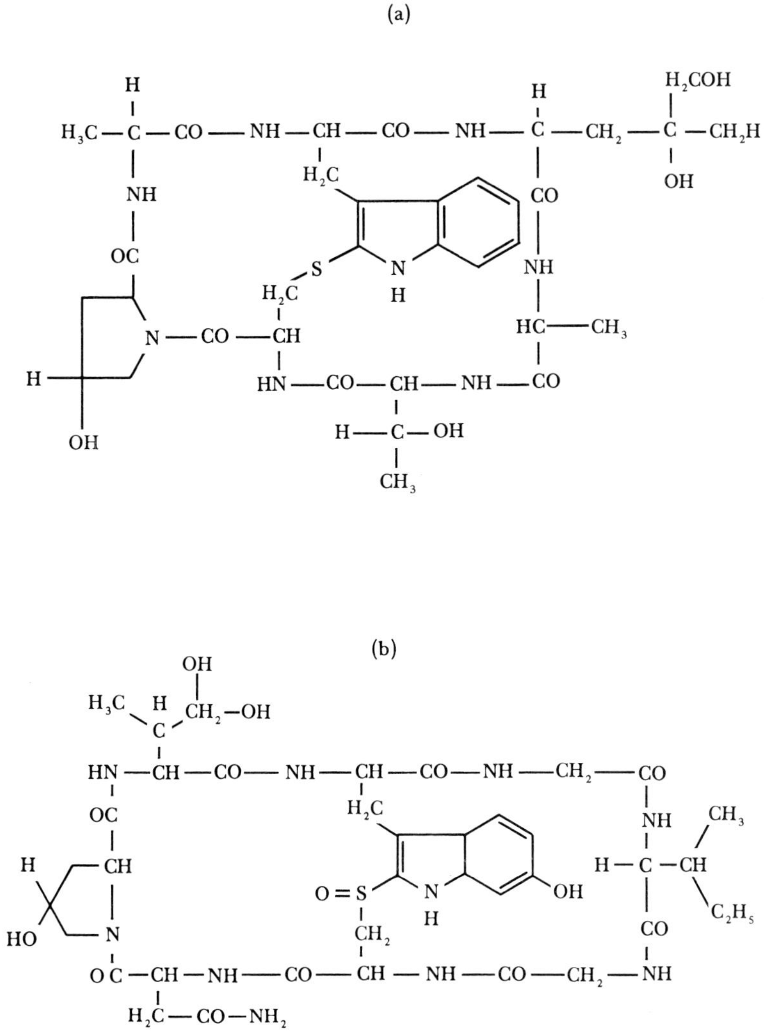

(a)

(b)

Fig. 8.1 Toxic compounds from *Amanita phalloides*: (a) phalloidin; (b) α-amanitin.

nucleoli are lost the cell ceases to function normally and rapidly dies. The major cause of *Amanita* poisoning is thus irreversible loss of hepatic and renal function.

It has recently been suggested that toxins such as phalloidin and α-amanitin do not exist as such within the fruit bodies of poisonous *Amanita* species, but that these cyclopeptides are merely fragments of a much more complex toxin. The application of more gentle biochemical methods of extraction to fruit bodies results in the preservation of fragile, molecular assemblages which are called myriamanins. These toxins are composed of a carbohydrate core that is surrounded by a number of cyclopeptide molecules which are bonded both to the core and to one another. The exact chemical nature of myriamanins and their toxicology are currently being explored.

Although *Amanita* poisoning in man is an exceedingly rare event, antidotes are being sought and two compounds, neither of which is yet fully medically approved, have on occasion been shown to be effective. One of these is cytochrome C, an enzyme involved in cellular respiration, and the other is thioctic acid, a small organic molecule. Their mode of action is not known but thioctic acid allows the repair of toxin-induced liver damage.

Fascinating though poisonous mushrooms and their toxins might be, they have at the moment relatively little medical and no economic significance. In contrast, toxins produced by some microfungi can, in certain situations, be extremely important, and such compounds have great significance in some areas of medicine and agriculture. The toxic properties of one such fungus, *Claviceps purpurea*, a plant-parasitic Ascomycete, have been known for centuries, but the insidious effects of others are only presently being revealed.

Fig. 8.2 Flowers of rye grass infected by *Claviceps purpurea*. Long, curved ergots are protruding from infected florets. (*Photograph by G. Woods.*)

Brief mention was made in the first chapter of the formation, structure and function of sclerotia. *Claviceps purpurea* is a sclerotium-forming species which causes a disease known as 'ergot' that is commonly found on many wild and cultivated grasses, but which occurs most frequently on rye. Ergot can affect over 200 grass species of the temperate zones of the world and this figure may represent only a fraction of those that are potentially susceptible. The parasite is confined to the flowers of the grass, and in late summer diseased inflorescences may easily be recognized by purplish-black, slightly curved sclerotia, or 'ergots', protruding from infected flowers, which have been prevented from setting seed (Fig. 8.2). In the autumn these sclerotia fall to the ground, where they lie until they germinate in the spring. During germination, each ergot gives rise to a number of thin stalks, each bearing at its tip a small, globose head (Fig. 8.3). Every head has, immersed within it, numerous flask-shaped chambers called

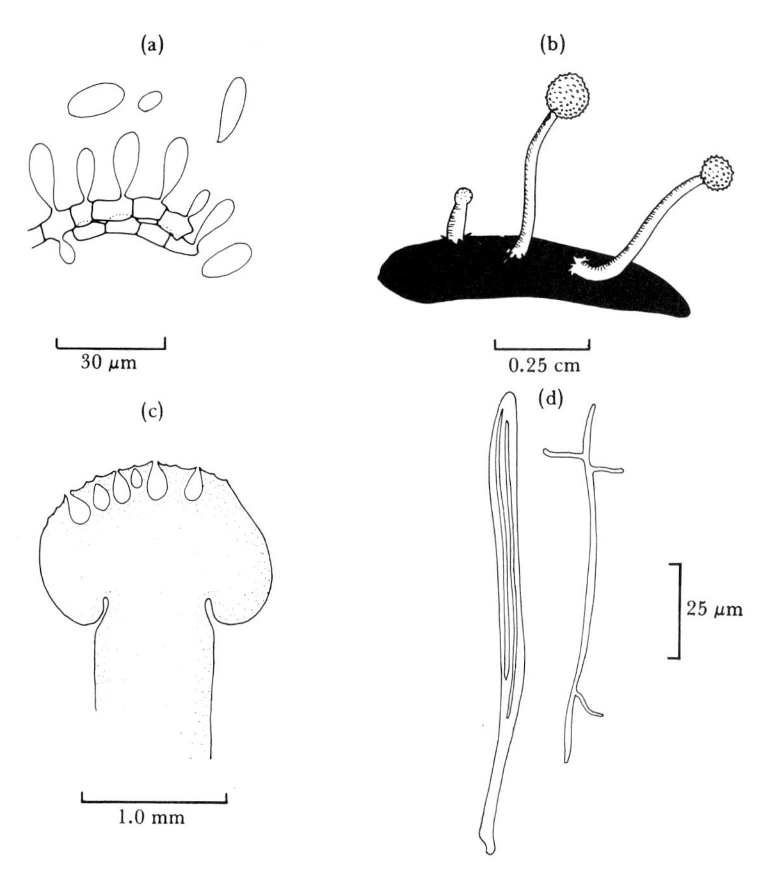

Fig. 8.3 Some stages from the life cycle of *Claviceps purpurea*: (a) conidia being formed on the surface hyphae of a developing ergot; (b) germinating ergot with stalked stromata; (c) vertical section through a stroma showing the flask-shaped perithecia; (d) a single ascus from a perithecium to show 2 of the 8 thread-like ascospores, and a germinating ascospore.

perithecia, each connected to the exterior by a minute pore. Inside the perithecia are large numbers of thread-like ascospores. The stalks grow towards the light, carrying their heads a short distance above the soil surface, and the ascospores are then forcibly ejected through the pores, shooting up to 5 cm into the air where they are caught up in wind currents and transported quite considerable distances. Ascospore dispersal takes place during the same period that grass flowers are opening and are projecting their feathery stigmas in readiness to receive pollen. The long, thin ascospores become entangled in the stigmas and germinate, their germling hyphae growing to the base of the ovary which then becomes penetrated. Hyphae grow rapidly within ovary tissue, gradually destroying it, and also cover the exterior of the ovary, where large numbers of unicellular conidia are produced (Fig. 8.3). These spores are formed in a sticky, sugar-rich 'honeydew' that oozes in droplets from infected flowers. Conidia may be spread to uninfected flowers on the same inflorescence, or to different inflorescences either by splashing rain or by insects which are attracted to the honeydew. On moving away from infected flowers such insects carry conidia with them on the head, body and legs. Spores reaching healthy flowers germinate, infect the ovaries and subsequently more conidia are formed, so that rapid spread of infection through a grass population is possible.

The ovary is converted quite slowly, from its base upwards, into a sclerotium as the hyphae colonize it and, as this process continues, the production of conidia and honeydew gradually declines and finally ceases. The ovary is transformed initially into a loose mass of mycelium, but as the sclerotium develops the hyphae harden and darken to the typical purple hue. Ergots are usually fully formed within infected flowers at about the same time that seeds are produced in healthy ones.

In grasses that are cultivated for their grain, ergot can reduce yields not only directly, through grain replacement, but also indirectly through sterility induced by an infected flower in neighbouring flowers. Sterility levels in individual flowers within an infected inflorescence can be as high as 20 percent. However, in addition to having important effects on grasses, ergots, if ingested, can also produce a disease in both man and animals known as 'ergotism'.

Ergots may be harvested with grain and so find their way into flour or animal feed, or may be ingested by grazing animals. The consumption by sheep or cattle of heavy doses leads to a characteristic degeneration of the small terminal arteries so that the tissues of the ears, tail and feet die and these organs are eventually shed. In less extreme cases lameness and abortion may occur. Ergots of other species of *Claviceps* may also produce disease symptoms in stock animals. For example, in Australia a common grass, *Paspalum*, is attacked by *C. paspali*, the ergots of which are particularly poisonous when they are only partly formed. Animals collapse, frequently at some distance from food or water, and may die. Less severe is the poisoning caused by ergots of *C. fusiformis*. This East African species attacks *Pennisetum typhoides*, the bulrush millet or pearl millet, which is grown extensively in Rhodesia for its grain. Pregnant sows fed on rations which contain ergotized millet give birth to piglets which are normal and healthy. However, these usually die of starvation within three days of birth. The mammary glands of the mother fail to develop normally during the last few weeks of pregnancy and this results in an almost total lack of milk, but apart from this the affected sows show no other signs of ill-health. Even in instances where as much as 2 percent of the grain is ergotized there is no lameness or necrosis of the extremities.

Ergotism in man has a long, well-documented history and usually is obviously associated with the consumption of ergotized rye flour. The first known written account of it dates from 857 AD when, in the Lower Rhine settlement of Kanten, the population was afflicted with a plague characterized by the appearance of swollen blisters and the loss of whole limbs. An outbreak is recorded for Paris in 945 AD, but here the major symptom was described as being 'fire', probably inflammation of the limbs. Large numbers of victims were apparently involved in these and subsequent outbreaks in Europe during the eleventh and twelfth centuries. Characteristically, the medieval mind discerned divine involvement in these occurrences and the disease went by the name of '*ignis sacer*'—holy fire. For some reason during the eleventh century the disease came to be be associated with St Anthony, and sufferers began to make pilgrimage to the Saint's relics at Vienne in France. The name of the disease then underwent mutation to 'St Anthony's Fire' and houses of the Order of St Anthony had red-painted walls as a symbol of the disease (Fig. 8.4). In Western Europe, after the middle of

Fig. 8.4 Woodcut from Gersdorff's treatise on military surgery published in 1535. A peasant who has lost his right foot through ergotism and whose left hand is enveloped in symbolic flames appeals for relief to St Anthony. The significance of the pig is not clear. (*From G. Barger, Ergot and Ergotism, 1931.*)

the eighteenth century, there was a gradual decrease in the frequency of outbreaks, possibly correlated with an increase in the use of wheat and the potato as staple, carbohydrate-rich crops. However, during that period one outbreak of ergotism changed the course of history, although perhaps only in a minor way. In 1722 Peter the Great marched an army against the Turks but a supply of contaminated bread rendered his cavalry helpless. To quote a contemporary account: '*Tout la cavalerie, qu'il avait menée à Astrakan, est ruinée, et ses finances sont en très mauvais état.*'

The United Kingdom has been relatively free from ergotism and only a single severe outbreak has been recorded. In Suffolk in 1762 a farmworker's family suffered the loss of feet and legs, probably due to ergotized wheat. However, a mild epidemic was discovered as recently as 1927 among the Jewish community in Manchester. Although symptoms were on the whole not very severe it seems that the victims had suffered from ergotism for months, if not years, through eating contaminated rye bread.

Fortunately, it is relatively easy to reduce or almost eliminate the incidence of ergotism in both man and animals by using simple agricultural techniques. If ergots are found in grain samples during screening then grain can either be destroyed or the ergots can be floated off from the grain in brine baths. If the cleansed grain is washed immediately the salt does not damage it. Where hay crops are affected, or are thought to have a good chance of becoming infected during the growing season, the grass should be cut early before ergots develop. Grass seed contaminated with ergots should not be used to establish new pastures as this may spread the disease due to the possible germination of the sclerotia in late spring. So that these simple control measures will succeed, vigilance is required, but despite this outbreaks of ergotism still occur from time to time. What is now looked on as a classic outbreak occurred as recently as 1951 in France at Pont St Esprit.

Bread made from contaminated flour was sold to the villagers and 6–48 hours later the first symptoms appeared. In the main these were feelings of depression or mild anguish, slight agitation, colicky pain and constriction of the pharynx. Some victims also had insomnia and gave off a disagreeable odour. Many recovered after this stage but for some the disease became more severe, with painful cramp, coldness of the hands and feet, and formication. The latter is a feeling that ants or other small animals are running under the skin. Finally, badly-affected people experienced an inability to stop talking, total insomnia and, towards each evening, hallucinations. Visions of animals and flames were the most common. Two victims threw themselves from windows, but in all there were only four deaths. There were, however, some long-term effects, particularly among people who normally drank an excess of alcohol each day. Ducks and other animals which had been fed on the bread also died, usually in convulsions, after short periods of eccentric behaviour.

The multiplicity of symptoms of ergot poisoning and their varying degrees of expression in different outbreaks requires some explanation. Modern studies on ergotism show that there are three broad clinical features. First, there are alimentary upsets such as diarrhoea, colicky abdominal pains and vomiting. Secondly, there are circulatory changes characterized by a decrease in the diameter of the blood vessels, particularly those supplying the extremities. Finally, there are neurological symptoms marked by headache, vertigo, convulsions and psychotic disturbances. Ergots contain large quantities of complex organic compounds known as ergoline alkaloids. Two of these, ergotamine and ergometrine

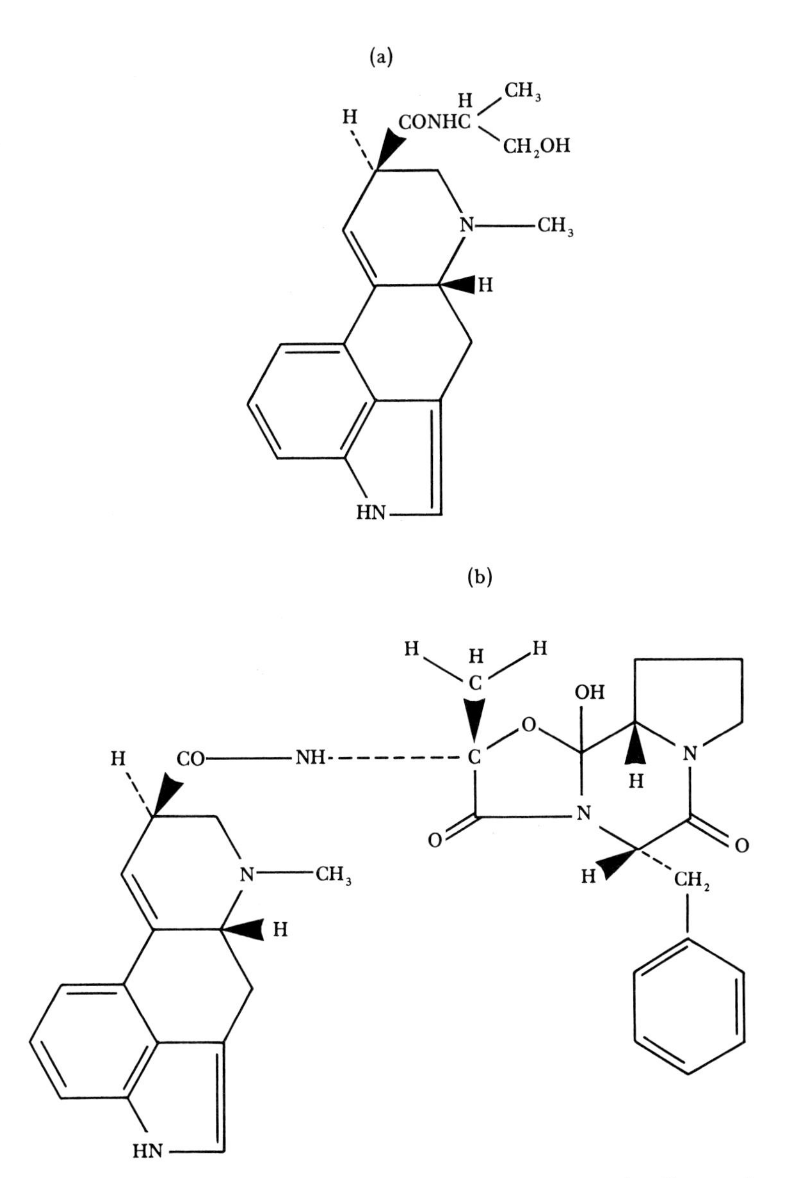

Fig. 8.5 Ergoline alkaloids from ergots: (a) ergometrine; (b) ergotamine. Ergometrine and the left-hand part of the ergotamine molecule closely resemble lysergic acid (see Fig. 8.6).

act on smooth muscle tissue, while two more, ergocristine and ergocristinine, affect the sympathetic nervous system (Fig. 8.5). The spectrum of symptoms and their severity will, therefore, depend on how much ergotized material has been ingested and on the relative proportions of these various alkaloids in the original ergots. The molecular structure of many ergot alkaloids is based on that of lysergic acid, whose derivatives have well-known hallucinogenic properties (Fig. 8.6). There is, however, no evidence that the various delusions and visions which have occurred in some cases of poisoning are due to ergot alkaloids, and their cause remains at the moment unexplained.

In addition to having the ability to produce disease symptoms, ergot derivatives, if correctly used, may also have beneficial effects. As early as the sixteenth century it was known that ergots could induce uterine contractions during childbirth, and so accelerate the process. Present uses of ergot alkaloids include reduction of bleeding during labour and treatment of migraine. Ergotamine tartrate

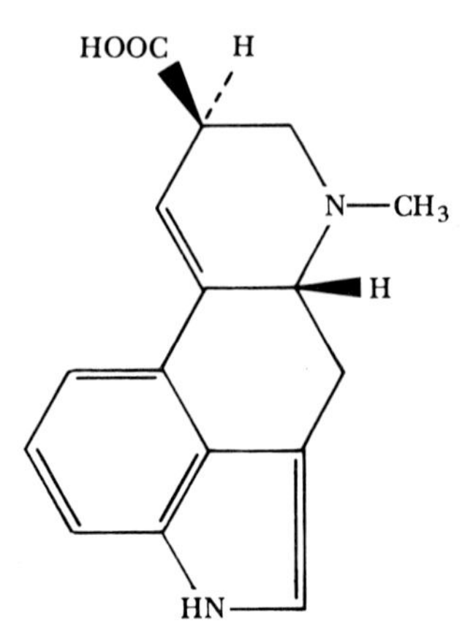

Fig. 8.6 Lysergic acid. The hallucinogenic drug LSD is synthetically derived from this naturally-occurring compound.

in small amounts, either taken orally or as suppositories, acts as a potent vasoconstrictor. In attacks of migraine it will reduce the diameter of abnormally dilated intracerebral vessels on the affected side of the head, thereby reducing blood flow and relieving pain. Unfortunately, a number of cases of ergotism have occurred due to the use of ergotamine in migraine treatment. Feelings of numbness, mainly in the legs and toes, vomiting and colic are common symptoms. On withdrawal of ergotamine the symptoms usually quickly disappear but, unfortunately, in a few cases misdiagnosis has led to amputation of the numbed limbs.

Toxicological hazards can also be created by fungi that are common spoilage organisms of foods and food crops. Normally, mouldy food would be destroyed, but in situations where there are adverse economic or social factors they may of necessity be consumed. Humans and animals may in this way ingest considerable quantities of contaminated grain and other feeds, or grazing animals may be pastured on mouldy fodder crops. They then become poisoned, and this kind of poisoning, mycotoxicosis, is caused by the growth of fungi on and within the food. During growth the fungi may produce their own toxins, mycotoxins, or may chemically modify some normally harmless compound present in the substrate so that it is transferred into a poisonous one. In addition, some food constituents, while being harmless themselves, can become harmful in the presence of fungal toxins without being modified chemically. Mycotoxicoses have distinctive characteristics. The diseases are not transmissible, drugs or antibiotics have little effect on the symptoms, and field outbreaks often occur seasonally or are linked to certain climatic conditions. Any outbreak can usually be associated with a specific food or feedstuff and examination of this normally reveals signs of fungal activity. The gross effects of eating contaminated foods have been known for some time, particularly those occurring in cattle and other crop animals, but the details of such poisonings were not investigated to any degree until relatively recently. For this reason their full importance was not realized, although now this field of science is one where great expansion is currently taking place.

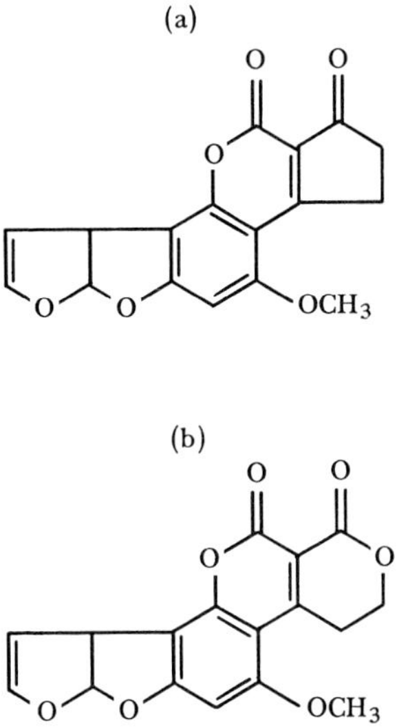

Fig. 8.7 Two aflatoxins from *Aspergillus flavus*; (a) aflatoxin B_1; (b) aflatoxin G_1.

Intensive studies on mycotoxins began in 1960–1 when outbreaks of disease occurred in a number of countries in poultry and also in trout reared in commercial fish hatcheries. One of the most prominent outbreaks was in the United Kingdom where 'turkey X disease' caused severe losses in turkey poults. Acute forms of the disease were characterized by lethargy, loss of appetite, weakness of the wings and finally death. Having eliminated communicable disease and various kinds of poisoning as being the cause of turkey X disease, it was found that all outbreaks shared a common factor and this was the utilization of groundnut meal in the feed. Coincidental outbreaks of disorders with similar symptoms in other farm animals, particularly ducks, were also linked to the use of groundnut meal. Subsequent investigations in many countries showed conclusively that such diseases were caused when mouldy groundnut meal, or other kinds of meal, were included in the diet. The fungus responsible was usually found to be *Aspergillus flavus* but other species of *Aspergillus* could produce the same effects.

Table 8.1 Some toxins produced by microfungi when growing on foodstuffs or fodder crops

Species	**Toxin**	**Foodstuff or Crop**
Penicillium	Luteoskyrin	Rice
	Rubratoxin	Maize
	Cyclopiazonic acid	Hay, grains, processed feeds
Fusarium	Zearalenone	Maize and barley
	Sporofusarin and poaefusarin	Overwintered cereal grains
	T-2 toxin	Pasture grasses, especially fescue
Stachybotrys	Stachybotrys toxin	Straw, hay
Pithomyces	Sporidesmin	Pasture grasses
Rhizoctonia	Slaframine	Red clover

These diseases are caused by aflatoxin, which is synthesized by *A. flavus* during its growth. This is not a single compound and at least 16 aflatoxins, all chemically related, have been discovered (Fig. 8.7). Aflatoxin causes the cells of the liver to break down and brings about blockage of the hepatic veins, while at the same time inducing the cells of the bile duct to proliferate. These effects do not seem to be reversible so that death results from loss of liver function. There is mounting evidence that aflatoxin acts by impairing RNA synthesis and perhaps also the synthesis of deoxyribose nucleic acid (DNA). Mammals, including primates, birds and fish, are all susceptible to aflatoxin, but details of its effects vary with the species and with the kinds of aflatoxin involved.

One property of aflatoxin which has aroused great interest is that with some animals it has a carcinogenic effect, producing cancerous tumours (hepatomas) of the liver. Although only circumstantial evidence is available, it has been suggested that aflatoxin may be one of the factors which determines the observed geographical distribution of liver cancer. High levels of primary liver cancer are found in Africa south of the Sahara. For example, in several regions 50 percent of all cancers found in the indigenous population are liver tumours, and this is 5–50 times the level of occurrence of hepatomas reported in the United States. However, the incidence of liver cancer in whites and non-whites in North

Effects	**Animal normally affected**
Fever, haemorrhage, liver damage (including cancer)	Man, cattle, frequently fatal
Loss of appetite and co-ordination haemorrhage, liver and kidney damage	Cattle, poultry, frequently fatal
Damage to liver, kidney, heart	Cattle, frequently fatal
Prolapse of the vagina, increase in uterus size	Female pigs, non-fatal
Generalized haemorrhage, fever, sepsis, bone-marrow exhaustion	Man, normally fatal
Reduction in blood flow to extremities, agalactia	Cattle, stock must be destroyed
Fever, absence of blood clotting, haemorrhage	Horses, normally fatal
Facial eczema	Sheep, non-fatal
Excessive salivation	Cattle, non-fatal

America is similar, so that the high African levels might be related to environmental rather than to genetic factors. This view is supported by the observation that the incidence of hepatomas in South African whites is equal to that in North American whites. Many environmental variables could possibly contribute, singly or together, to determine this pattern of tumour distribution, but it may be of some significance that high regional incidence of hepatomas seems to be related to tropical climate and to the prevalence of poverty or food scarcity. Food scarcity leads to the ingestion of contaminated foods and these may contain liver-damaging compounds, including aflatoxin. It is interesting to note that during screening tests for aflatoxin conducted in Uganda, 40 percent of the foods tested contained the toxin and 15 percent of peanut samples contained levels well in excess of the recommended safe limit.

Although it has by no means been proved that aflatoxin is a major factor in liver cancer, or other liver diseases in humans, the United Nations World Health Organization has recommended a tolerance limit for the toxin in foodstuffs of 30 parts per billion. In underdeveloped countries, where the necessary apparatus of quality control and legislation is absent, it is obviously impossible to apply this standard. It has been shown that wherever there are conditions of high temperature and humidity, which favour fungal growth, together with primitive storage and marketing conditions, there is a considerable frequency of high toxin levels in foods. *Aspergillus flavus* and other aflatoxin-producing species are ubiquitous and can grow on a wide variety of foodstuffs. Storage of the latter under conditions that preclude or reduce the incidence of fungal contamination reduces the risk of subsequent mycotoxicosis. If food or feed grains do become mouldy, they can be separated from healthy ones by using the appropriate methods. Aflatoxins can also be removed with chemical washes, or be inactivated by heating, but neither method is of commercial value.

Although aflatoxin is probably the most studied and most widely known mycotoxin, many more of equal importance are produced by a number of fungi, with new ones being continually discovered and investigated (Table 8.1). Several points might be noted with respect to them. Most of the fungi involved are extremely common saprotrophs which are found in decaying organic matter of all sorts, where they contribute beneficially to its breakdown; it is only when they occur 'out of place' and are consumed that they become harmful. Very few, if any, produce a single mycotoxin, but in a single fungus the toxic effects are brought about by a complex of compounds which are usually closely related. However, toxins produced by different fungi are commonly not closely related. The occurrence of natural diseases attributable to a mycotoxin in only certain species of animal is not due to a specific effect of the toxin on that particular animal, but simply reflects the fact that the animal is one which normally consumes a particular food. Most artificially-administered mycotoxins will cause disease in a wide range of mammals and birds.

The medical and agricultural importance of mycotoxins is quite clear but it should be remembered that mycotoxicoses are only reported if symptoms are severe enough to be immediately noticeable. It is impossible to estimate the extent to which long-term mycotoxicoses acting at less than severe levels contribute to human suffering, or to a loss in the productivity of crop animals, but it is likely to be large. Finally, it is ironic that the contemporary explosive interest in mycotoxins has brought to general notice one of the great and little known tragedies of the Second World War. During the years 1942–7 near-famine conditions in Russia forced people to scour the fields for unharvested grain that had spent the winter on the ground under snow cover and during this time had been attacked by the mould *Fusarium*. Its mycotoxins, sporofusarin and poaefusarin, produced a normally fatal disease known as alimentary toxic aleukia (Table 8.1). In some areas over 10 percent of the population were affected.

It is, perhaps, easy to over-dramatize the dangers of fungal toxins in food products, but in doing so it is also easy to obscure the greater importance of fungi as a source of food materials. Fungi can be extremely valuable in this respect and in the future are likely to have an increasingly important role in food production on a global scale.

9

Industrial protein and modified foods

'What a lot of these funguses there are about here!' remarked brother Tom presently. 'I can't see what use they are in the world.' ... 'I dessay they're sent for some wise purpose,' said Mr. Coombes.

H. G. Wells, *The Purple Pileus*, 1897

Fungi are indispensable in many familiar processes of food production and the important role of yeasts in baking, brewing and wine-making does not require elaboration here. Fungi are also used to condition a great number of foods so that they acquire attractive flavours and textures, the best-known examples being the noble cheeses, among which are Roquefort, Brie and Blue Stilton. In addition, many commercially-produced fungal enzymes are widely employed in food technology, for instance in large-scale milk clotting in cheese manufacture and in clarifying fruit juices before they are bottled or canned.

Although many such processes involving fungi have their roots deep in the past, it is fair to say that until relatively recently these organisms have been utilized mainly in the manufacture of palatable commodities rather than being exploited for the production of essential foods. For example, the preparation of potable alcohol and of leavened bread are probably the two most ancient fungus-based processes. It has been said that agricultural peoples were bound eventually to invent either beer or leavened bread, and having invented one the discovery of the other was an inevitable consequence, since they are very closely-related products derived from the use of closely-related yeast species. It is almost true that beer is basically liquid bread. Quite early after the founding of the first city-orientated cultures the production of both must have become established on a commercial basis, there being good evidence that this was the case in Ancient Egypt. During the reign of Rameses III the annual contribution of food offerings at the Temple of Amon in Thebes alone included 68,000 litres of beer and over 6 million loaves of bread. Considering the requirements of all the other temples, together with domestic consumption, it is clear that these needs could only have been met by large-scale brewing and baking. Much as we today still value potable alcohol and leavened bread they are not, however, essential to our nutrition. Although beer and wine contain high levels of vitamins, these can be obtained from other and cheaper foods or drinks, while the main contribution of yeast to breadmaking is through the production of carbon dioxide which inflates the dough. Unleavened bread is not as palatable but is quite as nutritious as that made with yeast.

A major essential food constituent is protein. It has been estimated by the United Nations Food and Agriculture Organization that the global demand for protein will have almost doubled by the end of this century, and it has been calculated that it will be impossible to satisfy this future requirement by means of systems based on conventional agricultural and fishery methods. One of the alternatives available is to use fungi directly as protein producers, and more and more attention is being drawn to these organisms as potential food sources.

The direct use of fungi as a major source of protein seems at first sight to be an unlikely prospect, but this is because the tendency is to think in terms of growing and cropping mushrooms. The productivity of wild mushrooms and toadstools is surprisingly high, particularly in some woodland habitats, where annually up to 500,000 fruit bodies per hectare may appear (Table 9.1). For a great many reasons it is, however, impossible to make use of these as a worthwhile crop. A proportion of them are inedible species, the fruit bodies are short-lived

Table 9.1 Annual productivity of Basidiomycete fruit bodies in natural habitats (After Hering, 1966)

Habitat	Country	Fresh weight yield (kg per hectare)
Pine wood	United Kingdom	265–460
	United States	150–260
	Finland	82–302
Oak and Beech	United Kingdom	13–95
wood	Hungary	7–160
Grassland	United Kingdom	1–10

and appear mainly in the period August to November, and the yield depends on soil temperature and water content, which are impossible to control.

Cultivated mushrooms can, on the other hand, be intensively grown on a large scale under carefully controlled conditions, and this is done in many parts of the world. A number of different species are used, such as *Lentinus edodes* and *Armillaria matsutake* in China or Japan, *Volvaria* species in India, and *Agaricus bisporus* in Europe and the United States. Of these it is *A. bisporus* which is grown on the largest scale, and current estimates indicate an annual world production in excess of 250,000 tons, which represents about 10,000 tons of dried protein. This is only a minute proportion of the world's annual total protein requirement, which is at present over 40 million tons, but it is important to remember that it is the future production potential, not the present, which is important with regard to mushrooms as a protein source. Sufficient information is available to allow an assessment of this potential to be made. *Agaricus bisporus* fruit bodies consist of 93 percent water and their protein content is very low, mushrooms having the same food value as vegetables such as carrots and broccoli. However, it is possible to grow several heavy crops of mushrooms in a year, so that under conditions of intensive cultivation incredibly high yields can be obtained. These high yields compensate for the low protein content and allow mushrooms to be compared more than favourably with other crops in terms of yield per unit area.

For instance, cereals give annual yields of 3,000–6,000 kg per hectare but mushrooms may give up to 2 *million* kg per hectare, while on an area basis they are vastly more valuable sources of protein than either beef cattle or fish (Table 9.2). Despite the figures, mushroom growing cannot solve the world protein problem because of the operation of a number of negative factors, the major one being the cost of the protein produced by these methods which is at the moment many times the cost of protein obtained from conventional agricultural practices. Mushroom cultivation, particularly at the picking stage, is labour-intensive and also relies heavily on the availability of organic substrates suitable for growing the fungi. For *A. bisporus*, animal manure containing straw is widely used, and it is inconceivable that a sufficient supply of this could be maintained on a large enough scale to permit mushrooms to be grown as a nutritionally significant crop. However, it has been suggested that the by-products of a wide range of industrial processes—e.g. molasses, tree bark and municipal refuse—could be used as substrates for mushroom growing. It is therefore still not unlikely that the development of novel methods of mushroom culture and harvesting, together with a continued rise in the cost of cereal and animal proteins, will at some time in the future make mushroom growing for protein an economic possibility.

Table 9.2 Yields of dry protein per unit area utilized for farming beef, fish and *Agaricus bisporus*

	Approximate annual yield dry protein (kg per hectare)
Beef cattle, conventional agriculture	78
Fish, intensive pond rearing	675
A. bisporus, conventional methods	65,000

Current investigation of the value of fungi as direct protein sources is being concentrated not on mushrooms but on microfungi, and in particular the yeasts. It is probable that within the next 20 years yeast protein, termed single-cell protein, will become a major component of the diet of both man and his livestock. The basic principle for the production of single-cell protein is very simple. The yeast is grown on a medium which is rich in carbon and the carbon is then used by the fungus to make carbon skeletons for proteins within its cells. Having made the conversion of the substrate to internal protein, the yeast cells can then be harvested either to be used as they are or to have their protein and other cell constituents extracted for processing in various ways. Our major sources of protein come from crop plants and livestock, and these build up their proteins in the same way as yeasts except that plants use atmospheric carbon dioxide as their sole carbon source and animals use complex carbon compounds ultimately derived from plant material.

There are many advantages in using yeast cultures rather than plants or animals as a protein source. They grow extremely rapidly and, under favourable conditions, can double their mass in a very short time in comparison with other organisms. The protein content of yeast cells is more than 40 percent, which means that the amount of protein per unit of starting material produced in a standard time is unbelievably high (Table 9.3). In addition, production plants for microbial protein require very little land space and the process can be operated continuously, with substrate being added at one end of the manufacturing system and dried yeast cells leaving it at the other.

The successful, large-scale manufacture of single-cell protein can only be achieved if a number of important criteria are fulfilled. First, the organisms used must have high growth rates at high cell density, and their cells must have a suitable protein-rich composition. Secondly, the substrates on which they are grown must be easily available in large quantities and should preferably be relatively cheap. Finally, the end-product, whether it be dried yeast cells or cell extracts, must be nutritionally available to the eater and be non-toxic.

A number of yeasts, in particular *Torula* and *Candida* species, are available which satisfy the requirements for high growth rates at high cell density, and these also have a high protein content. Cheap and widely-available substrates that are rich in carbon compounds can also be obtained on a large scale. It is interesting to note that some of these are waste products which present disposal

Table 9.3 Comparison of time taken for different organisms to double their mass under optimum conditions and comparison of protein synthesis rates per unit starting mass (After Vilenchich and Akhtar, 1971, and other sources)

	Time taken to double mass	**Protein synthesized, (kg per 24 hours per kg of starting mass)**
Beef cattle	4–10 weeks	0·001
Pigs	4–6 weeks	—
Poultry	2–4 weeks	—
Yeasts	20–120 minutes	1,000
Soya bean plants	—	0·1

problems to various industries. Wood-pulp sulphite liquors from the paper-making industry present serious pollution problems when passed as effluents into rivers, but can serve as good substrates for yeast growth. Similarly, molasses from sugar refining and whey from milk processing are difficult to dispose of, yet could provide carbohydrate-rich substrates for single-cell protein production. The large-scale use of yeasts would thus dispose of two pressing problems, a world shortage of protein and certain forms of environmental pollution.

One substrate that yeasts can utilize is crude oil. It has been calculated that if 10 percent of the present production of crude oil were diverted to the manufacture of single-cell protein, the 50 million tons of protein so produced would nearly satisfy the projected world protein demand of 65 million tons by the year 2000. Whether the use of crude oil for this purpose on such a scale is feasible, or even desirable, in the context of the current trend towards conservation of fossil fuel reserves is not at present clear. Some fungi can also utilize natural gas, essentially a mixture of methane and ethane, for growth, and these species are attracting a great deal of interest. The great virtues of using natural gas for protein production are, first, that it is very cheap and, secondly, that many underdeveloped or protein-starved countries have large, untapped reserves of natural gas. However, at the moment the organisms which have been found to utilize this carbon source have low cell yields and relatively long doubling times, so that their potential as protein sources appears to be low.

A fundamental problem in the production of single-cell protein is its cost, and at the moment at least some forms of it compare unfavourably with many protein-rich foods that are produced by conventional methods (Table 9.4). However, with climbing protein demand plus expansion and increased efficiency in yeast-growing technology, this situation will inevitably be resolved. Commercial plants are even now coming into operation for the production of protein from yeast grown on *n*-alkanes (middle-distillate compounds from petroleum refining) and the projected annual yield for one such concern in the United Kingdom is 100,000 tons.

Initially single-cell protein will be for animal consumption, and trials in 1974 have shown that alkane-derived protein can be used as a substitute for milk, fish and soya bean protein at dietary levels of about 30 per cent in feed for piglets and broiler fowl. It can also be partially substituted for skimmed milk fed to calves. Careful testing of this kind is necessary since single-cell protein can be toxic and may also be deficient in some essential amino acids, or have amino

Table 9.4 Relative prices per unit weight of currently available protein-rich products (After Vilenchich and Akhtar, 1971)

Commodity	**Relative price per unit weight**
Cotton seed flour	1
Peanut flour	1·2
Soya bean flour	1·2
Skim milk powder	4·1
Yeast cells	3·6

acids present in the wrong proportions. If used at too high a level in animal feeds there could be deleterious results. Single-cell protein for human consumption is likely in the near future, although this will probably require more careful treatment than that for animals and may therefore be relatively more expensive.

As well as mushroom growing and single-cell protein production there are other direct methods for obtaining fungal protein which may soon assume some importance in underdeveloped countries. There are many regions of the world where, paradoxically, good crops of food plants can be easily grown but where, at the same time, there is widespread semi-starvation. This situation can occur if the crop is at once a poor source of protein and comprises a major part of the normal food intake. For example, cassava or tapioca (*Manihot esculenta*) is extensively cultivated in parts of Africa, Asia and South America, where high yields of its starch-producing tubers are usually obtained. It can be grown economically and used both as a staple foodstuff and as a source of industrial starch. When it is consumed as a major component of the diet malnutrition and ill-health result. The tissues of the tuber contain less than 0·5 percent protein and are, in addition, slightly toxic, so that not only is there starvation but also a number of pathological syndromes due to the combined effects of malnutrition and toxicity. However, processes have been developed which utilize fungi to bring about an increase in the protein content of cassava-based foods.

As might be expected, the starchy tuber tissues provide a suitable energy-rich substrate for the growth of fungi, but unfortunately at the same time they are deficient in available nitrogen compounds so that fungal development is limited by lack of these essential nutrients. This lack of nitrogen comes about because the majority of nitrogen-rich compounds in the tuber are located in or close to its skin, and since this also contains extremely bitter compounds it must be removed before the tuber tissue is processed. The deficiency must be made good if suitable fungal growth is to be obtained and this can be done simply and cheaply, for instance by using urea. Briefly, the process involves crushing the flesh of the tuber which is then extruded mechanically and cut into cylindrical pellets. These are then placed in trays and are inoculated with a spore suspension from a suitable fungus, usually a species of the phycomycetous genus *Rhizopus*. After a few days incubation the fungus has grown and permeated the starch-rich substrate and all the material, both tuber tissues and fungal mycelium, can then be dried and stored until required. Using such methods the protein level of the cassava product has been raised to over 3 percent. A great advantage of this kind of modification of food is that it does not depend on advanced technology for success and small, local production centres can easily be established in rural areas.

Even where protein-rich foodstuffs are abundant, fungal modification of these can lead to desirable physical, nutritional and organoleptic changes. For instance, in the Far East, where the diet is often vegetarian or near-vegetarian, fungal fermentation of food has been a common practice for centuries. Among the principal raw materials for such processes is the soya bean which, while being a notably good source of protein, is not easily digested even after prolonged cooking. Growth of *Aspergillus* species on suitably prepared bean tissues can result in a wide range of appealing and digestible solid foods, drinks and food flavourings, among them being soy sauce. The protein content of the original material is not significantly enhanced, and may even be depleted, but this is unimportant in comparison with the overall improvement in palatability.

A time can be envisaged at which a proportion, if not the whole, of our diet may be derived directly or indirectly from fungi. Even the material in which this food will be wrapped may be a fungal product. *Pullularia pullulans* is a yeast-like member of the Fungi Imperfecti which, when grown in liquid culture under suitable conditions, produces copious amounts of pullulan. This compound is a complex carbohydrate that forms a thick, mucilaginous sheath around each fungal cell. It can be collected and dried to form a white, tasteless, odourless powder, which can then be used as a raw material to produce transparent wrapping films of various thicknesses. Such wrappings are highly impermeable to oxygen and are ideal for the packaging of foods. In addition, they are soluble in water and, since pullulan is tasteless, need not be removed first if the food product is to be placed in water for cooking. A Japanese company has already begun its commercial production.

10
Magic mushrooms and hallucinogenic drugs

Then the Big-Raven said, 'Let the Agaric remain on earth, and let my children see what it will show them.'
Traditional legend, Koryak tribe, Kamchatka, Siberia

The fruit bodies of Basidiomycetes have frequently been, and for the most part still are, objects of misconception and superstitious dislike. It is not difficult to see why this should be since, quite apart from the deadly nature of some of them, they have physical characteristics that many find unattractive. They are normally associated with putrescence and decay, are often slimy and of an unpleasant colour and are sometimes attended by a fetid smell. It is obvious how a multitude of beliefs based on these characters could have arisen in the past, particularly when it is remembered that physical corruption and evil tend to be automatically associated in men's minds. Three strange attributes characteristic of, and almost exclusive to, Basidiomycete fruit bodies are the ability to form 'fairy rings', the property of luminescence and the possession of complex biochemical compounds that are capable of inducing hallucinations if ingested. Each of these attributes has given rise to beliefs which spring from their apparently supernatural origin. In the case of the hallucinogenic fungi there is convincing evidence that widespread religious practices have been based on their controlled use.

When growing in pastures, on golf courses or lawns, basidiomycetes frequently produce their fruit bodies in conspicuous and remarkably regular rings which can vary in diameter from a few centimetres to several metres. Sometimes, but by no means always, the fungi have a marked effect on the vegetation in which they are growing so that the rings have a rather complex structure (Fig. 10.1). In such cases, at the circumference of the ring, where fruit bodies are being produced, the vegetation tends to be taller and a deeper green than that immediately outside the ring. Interior to this zone of enhanced growth, the whole area enclosed by the ring may consist of bare earth, particularly if the diameter of the ring is relatively small. If it is relatively large then the central part of the bare area may be colonized by plants. Immediately under the dark-green zone, and for a little distance interior to it, the soil is permeated to a depth of several centimetres by a dense mycelial weft of the fungus. No mycelium is found outside the ring nor in the regions near its centre (Fig. 10.2).

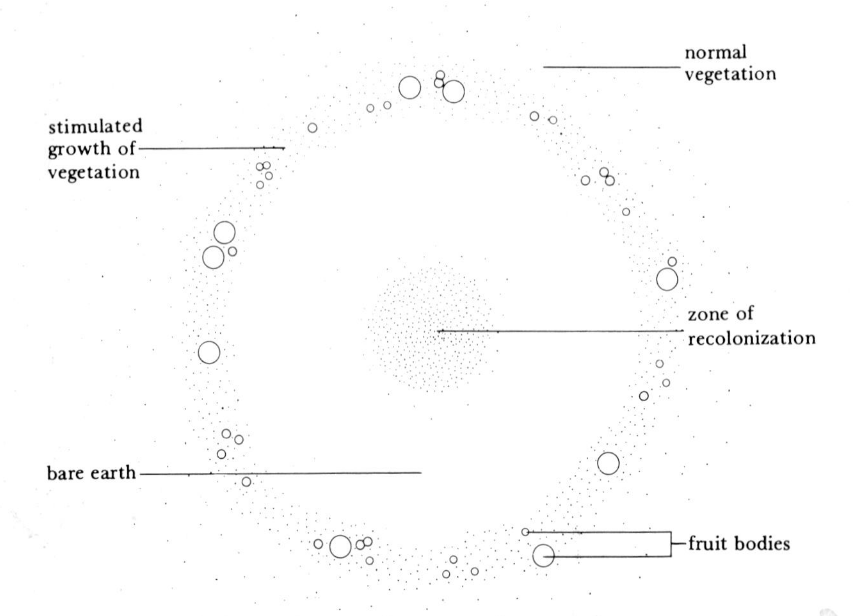

Fig. 10.1 Diagram of a fairy ring viewed from above showing the distribution of fruit bodies and the effects of fungal growth on the vegetation.

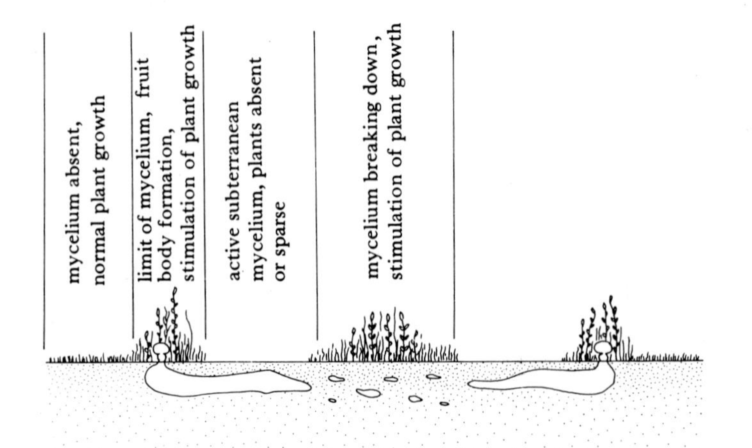

Fig. 10.2 Diagram of a section through a fairy ring showing the distribution of the subterranean mycelium and its effect on the vegetation.

In the past, the regularity of shape of the rings together with the presence of a bare central area were explained in a number of ways. The supernatural was often invoked so that the circles became the abode or dancing floor of fairies or toads and to step into one was to invite misfortune. Later, sceptics attributed them to the action of lightning or to large or small animals moving in circles and manuring the ground. Few attempts were made to explain the increase in size of particular rings from year to year, presumably because this growth could not easily be related to the activities of the supposed causes of the rings.

The underlying reasons for fairy ring formation are extremely simple. Mention was made in the first chapter of the circular growth of fungal colonies on or in a solid substrate. There is no reason why such circular growth should not occur in nature, provided it is not impeded physically and that sections of the colony are not destroyed. Such circular growth is clearly visible when various foods or fabrics are attacked by fungi and when lichens grow without interference from other organisms. A fairy ring is simply the visible manifestation of a subterranean fungal colony growing outwards through the soil from its point of origin. When small, the colony is a disc, but as it increases in size the mycelium comprising the central regions dies as nutrients within the soil become exhausted, so that the colony then becomes an annulus. Fruit bodies are only formed at the outer rim of the annulus and so mark its limits. Outward growth of the annular colony is maintained and as new mycelium pushes outwards the old mycelium behind it is continuously dying (Fig. 10.2). It is a simple matter to measure the growth rates of rings and from these to compute their age. The rate of advance of the mycelium varies from year to year, depending on prevailing weather conditions, but rates have been found to be on average between 12–30 cm per year. Rings up to 700 years old have been described although these may not have remained complete for various reasons.

Two further ring characteristics must be explained, first the bareness of the centre in older rings and secondly the stimulation of the growth of vegetation at the circumference. The mycelium of the colony derives many of its nutrients from the organic matter provided by dead plant material. It also obtains nutrients and water directly from the non-organic components of the soil and is thus in competition for these with roots of the plants that are living directly over the mycelium. These plants therefore experience severe water stress and nutrient deficiency, principally nitrogen deficiency, and die. In addition, mycelial growth is so abundant that the physical structure of the soil may be altered due to hyphae blocking the soil pores. Aeration is reduced and root growth ceases, thereby increasing the other stresses to which the plants are subjected. In this way bare areas are initiated within small rings or are maintained within larger rings. It should be emphasized that not all fairy ring fungi can compete successfully with plant roots in this way in all situations, and in these cases the central areas of the ring remain covered with vegetation. Eventually the central mycelium will itself die, but before it does so nutrients contained within it are moved through its hyphae to the young mycelium at the colony margin. This means that although the mycelium in the soil of the central zone is now dead its nutrients are not returned to the soil. This final impoverishment of the soil results in a level of fertility too low to support the growth of plant species entering it in the form of air-borne seeds. This condition prevails for some time but eventually the soil gradually becomes re-colonized.

Stimulation of vegetation at the ring margin is more difficult to explain but the most obvious basis for the phenomenon may be a nutritional one. The darker green colour implies that there is abundant nitrogen available to these plants. The fungus, through its saprotrophic activities, could be releasing nitrogen compounds formerly locked up in particles of organic matter and some of these compounds would then be absorbed by the roots. Alternatively, or simultaneously, the fungus could be producing hormones which promote plant growth. Many fungi synthesize these materials and release them into the substrates in which they are growing. If the fungus does promote conditions of nitrogen abundance for the plants then this is, of course, only temporary and there is no permanent advantage to the plants as they are very soon adversely affected by the onset of nutrient starvation.

Fig. 10.3 Shadow photograph of a maple leaf made by means of the light emitted from a single fruit body of *Pleurotus japonicus*. (*From A. H. R. Buller, Researches on Fungi, Vol. III, 1924.*)

Bioluminescence, the ability of an organism to produce visible light in the dark, is widespread among bacteria, plants and animals. Many fungi can also luminesce and almost without exception they are Basidiomycete species; the fruit body, the non-fruiting mycelium, or both may be luminous, depending on the species (Fig. 10.3). The luminosity of decaying wood has been known since ancient times and must have been obvious to primitive man. The glow from rotting tree roots seems to have given rise to beliefs in the magical power of such trees through the equation of light and power, and the effect of these inexplicable lights on superstitious minds must have been profound. Glowing roots are mentioned in the seventh century epic poem *Beowulf* in a description of the mere in which the monster Grendel's mother lives. '. . . over it hang hoary groves overshadowing the water. There each night a baleful wonder can be seen, a fire upon the flood.' At some time man's dread of this phenomenon was overcome, and there are numerous references to the use in the past of luminescent, rotting wood to mark pathways through forests. This kind of use has parallels in some primitive societies where luminous fruit bodies may be used as ornaments or as path-markers at night, and it is said that during the 1914–18 war soldiers in the trenches wore pieces of rotting wood on their helmets and rifle foresights to avoid collisions in darkness.

The glowing of rotten wood and the luminosity of fruit bodies have the same basis since wood decay is normally brought about by the mycelium of basidiomycetous fungi. In the tropics, species with luminous fruit bodies are much more common than in temperate regions, but the mycelium of temperate zone fungi is often luminous, although the fruit bodies are not. One such fungus is *Armillaria mellea* and the wood which it permeates is luminous, as is its mycelium when it is grown in pure culture on a sterile medium.

Although a great deal of research has been carried out on bioluminescence, particularly that shown by bacteria or insects such as fireflies, very little is known about the phenomenon in fungi. The light emitted is white or blue-green and is thus not monochromatic. The few wavelength measurements that have been made indicate that its constituent colours extend from orange to blue. Its intensity can obviously vary widely, but accurate estimations of the maximum intensities possible have not been made. It is certain though that even at its best the luminosity of fungi is not nearly as great as that exhibited by other organisms, although it can still be strong enough to literally read a book by and can be perceived at distances up to 30 m. If one nineteenth century observation is to be relied upon the light from a single fruit body may, exceptionally, be detected by the eye at 1,000 m.

The intensity and persistence of luminosity are modified by environmental conditions and in nature temperature and water availability are major factors. At temperatures close to or below freezing luminosity diminishes or ceases altogether, while too high a temperature has the same effect, the optimum range being approximately 10–25°C. Desiccation also reduces luminosity but this is restored on re-wetting. The first scientific experiment on luminosity was carried out by the famous chemist Robert Boyle in 1667 who showed by using a piece of rotted wood that light emission ceased in a vacuum. When air was re-admitted to the evacuated chamber luminosity was restored and was even a little brighter than before. These early observations have been confirmed, and the extra luminescence so produced is released in a pulse lasting about 20 seconds. If air is replaced by inert gases then a similar diminution in luminosity occurs. Anaesthetics such as chloroform and ether also adversely affect luminosity but recovery takes place on restoration of normal conditions.

In attempting to find an explanation for fungal luminescence there are three important factors to consider. First, it is a property of living tissues, since those factors that reduce it are also harmful to cell function. Secondly, it is a continuous process that carries on both by night and by day (although in daylight it is of course not visible) and is therefore independent of a light stimulus. In this respect it is very different from fluorescence with which it should not be confused. Fluorescence occurs when light energy is absorbed by a system and is then re-emitted at a lower energy level, i.e. at a longer wavelength, so that if no light is supplied then no light can subsequently be emitted. Finally, light is produced without the simultaneous production of heat.

Early attempts to elucidate the mechanism of luminescence in fungi involved attempts to obtain luminous juices from them. These attempts invariably failed but they showed that intact cells are necessary for the production of light. For example, luminescence is lost if a luminous fruit body is crushed or dried and powdered. If, however, the complete fruit body is merely dried, its luminosity is only lost temporarily and can be restored by moistening. More recent attempts have been based on those methods used to study the mechanisms of luminescence in bacteria and animals. Such methods were first used in the late nineteenth century to investigate the luminosity of a mollusc, *Pholas dactylus*. An aqueous extract of the mollusc was made using cold water. The preparation emitted light for a short time and then ceased to do so. An aqueous extract was then made in hot water and this extract was immediately cooled. It did not luminesce at all but when it was added to the cold-water extract luminescence occurred. This demonstrated that the bioluminescent reaction in the mollusc involved the participation of two separate entities. The cold-water extract contained an enzyme 'luciferase' and the hot-water extract a compound 'luciferin' on which luciferase acted to produce luminescence. Luciferase was inactivated by heating, so that the hot-water extracts were free of it. On the other hand, cold-water extracts contained both compounds but only luminesced until all the luciferin present had been used up by the luciferase. Luciferin–luciferase systems are probably the basis of all luminescence phenomena but the chemical nature of luciferin differs in different organisms, as do the details of the chemical reactions involved.

Fungal luciferin and luciferase have been extracted from Basidiomycetes in a crude form using modifications of the hot- and cold-water extraction methods, although the fungal luciferase system seems to be very unstable. Light emission occurs when the two preparations are mixed together and, as is the case in nature, oxygen is essential for the chemical reaction to take place. Luminescence does not reach maximum intensity immediately after mixing but there is a lag of 2–3 minutes before this occurs. The reaction takes place in more than one step and involves at least two concurrent reactions. Luciferin probably exists in an oxidized form and in the first reaction becomes reduced in the presence of a reducing compound, diphosphopyridine nucleotide (DPNH), and an unidentified enzyme. The reduced luciferin then reacts with oxygen and luciferase, and light is emitted.

Bioluminescence resembles respiration in that both processes liberate energy and both require oxygen. In the case of respiration the energy liberated is in small units which are used to help drive essential biochemical processes. In bioluminescence the energy is in some way prevented from being broken up into such useful small units and is liberated in large units in the form of visible light. It is, therefore, presumably a biologically wasteful process. This raises the question as to why fungal mycelia and fruit bodies luminesce. It has been suggested that night-flying insects may be attracted in this way and alight, pick up spores and then fly away to disseminate them. This is, however, an unlikely function for luminescence, particularly as those parts of the fungus that are most luminous are not spore-bearing and are commonly immersed within rotting wood or other substrates. It would seem that the luminosity of fungal structures serves no function at all, although at some point in evolutionary time it may have conferred a selective advantage. It has already been pointed out that primitive life forms arose on earth in atmospheric conditions very different fron those obtaining today, oxygen levels being extremely low, so that their metabolism had to be anaerobic. As life evolved, the oxygen content of the environment gradually rose, but oxygen would have been toxic to anaerobic life forms unless they possessed mechanisms to rid themselves of it. It is possible that one way in which they did this was to reduce the oxygen to water with the aid of an organic reducing compound. This process would have required the release of energy and, if the units of released energy were relatively large, there would have been light emission. Bioluminescence may therefore be merely a manifestation of energy-liberating mechanisms within the cell that are designed to remove oxygen from it. These systems are, of course, now non-essential and are redundant processes inherited from the ancestors of present-day fungi.

It is more difficult to explain why it is only the Basidiomycetes and a very few species outside this group that exhibit luminescence. This situation obviously might be due to lack of critical examination of other fungi. It might also be explained in terms of other fungi losing this capability in evolutionary time, yet the Basidiomycetes are thought to represent a pinnacle in fungal evolution, and for this reason might be expected to have shed any redundant biochemical mechanisms. The only thing that is certain is that in some basidiomycetous species the ability to luminesce is genetically determined. The mycelium arising from single-spore isolates—and therefore genetically uniform material—of a single species may be luminous or non-luminous. If such luminous and non-luminous mycelia are allowed to fuse the resulting mycelium is luminous, showing that the character determining luminosity is dominant over that determining non-luminosity. The implication of this in evolutionary terms is not clear. What is clear is that luminescence in fungi would repay a great deal of further study.

In addition to being objects of a range of superstitions, there is abundant evidence that the fruit bodies of higher fungi have played a major role in some long-standing and widespread religions. These religions are now for the most part dead, but are sometimes echoed by persisting ritualistic practices involving the present-day use of the fungi on which they were centred. Studies of these rituals, together with comparative investigations of oral or written accounts relating to the original religious beliefs, have yielded results that have caused a great deal of excitement and sometimes controversy. This excitement has not been confined to the field of mycology but has also involved medicine, archaeology, anthropology and Christian theology. All fungus-based religions, irrespective of their geographical location, made use of fruit bodies containing psycho-active substances which induced hallucinations when they were ingested. It was believed that, through the use of such hallucinogenic mushrooms, it was possible to gain knowledge of a God's will or to prognosticate the future, and in some religions the basic concept of a particular deity or pantheon may have arisen directly from fungus-induced visions. In purely mycological terms, interest in hallucinogenic fungi has been focused on two major areas: first, the identification of the species used in extinct or relict religious rituals and, secondly, the determination of the chemical compounds responsible for their observed psychological effects. Probably a great number of hallucinogenic species exist, but detailed information is available for only a few, two of which have been most intensively studied, *Amanita muscaria* and *Psilocybe mexicana*.

Amanita muscaria is widespread in north-temperate zones of both the Old and the New World and is probably the fungus whose appearance is most familiar to the layman. Its prominent, bright-red cap covered with white patches and its white, ringed stem arising from a large cup-shaped volva are constantly depicted in mushroom motifs (Fig. 10.4). Whenever a toadstool is required in a decoration or as an illustration in a children's story it is inevitably, and inexplicably, *A. muscaria* that is used. It is commonly referred to as the 'Fly-Agaric' since in medieval times it was used, broken up in milk, to immobilize flies. When ingested, *A. muscaria* produces several effects, the number and intensity of which vary with the individual. A single individual eating the fungus at different times may suffer different effects, no common and consistent experience being apparently obtained from the use of this species. After ingestion, there is a short incubation period of about an hour after which an involuntary twitching and trembling of the limbs occur. This is followed by the onset of a state of euphoria or excitement, during which visions of the supernatural may be obtained, and these hallucinations may be accompanied by violent physical activity. Finally, there follows exhaustion and a comatose sleep. Recovery is rapid but there is often no recollection of the hallucinations, and it is only during the period of active narcosis that the nature of any visions can be communicated to others.

Fig. 10.4 Fruit bodies of the Fly Agaric, *Amanita muscaria*. (*Photograph by M. W. Storey*.)

Regular ritualistic use of *A. muscaria* within recent times has been restricted to two major centres located in the extreme west and extreme north-east of Siberia, although there is some reason to believe that it has been used by the European Lapps. It is used by the men only and is taken dry or mixed with water, milk, or the juice of various berries. The active principles are excreted unaltered in the urine so that this may be drunk after recovery from narcosis to produce renewed symptoms. In tribes which use this fungus the general belief is that, while under its influence, a person will obtain messages from the spirits residing within the fruit body.

There is good evidence to suggest that in the past minor cults and widespread religions existed which were based on the use of the Fly-Agaric (Fig. 10.5). In these religions the fungus was perhaps forbidden to all but members of a priest class who had specialist knowledge of the correct way to utilize its powers, and who passed on divine messages to an audience while under its influence. However, to be able confidently to associate *A. muscaria* with a particular religion is often difficult since written sources may be lacking. Even where they do exist the fungus

---

Fig. 10.5 Distribution of known and putative cults centred on the use of hallucinogenic mushrooms.

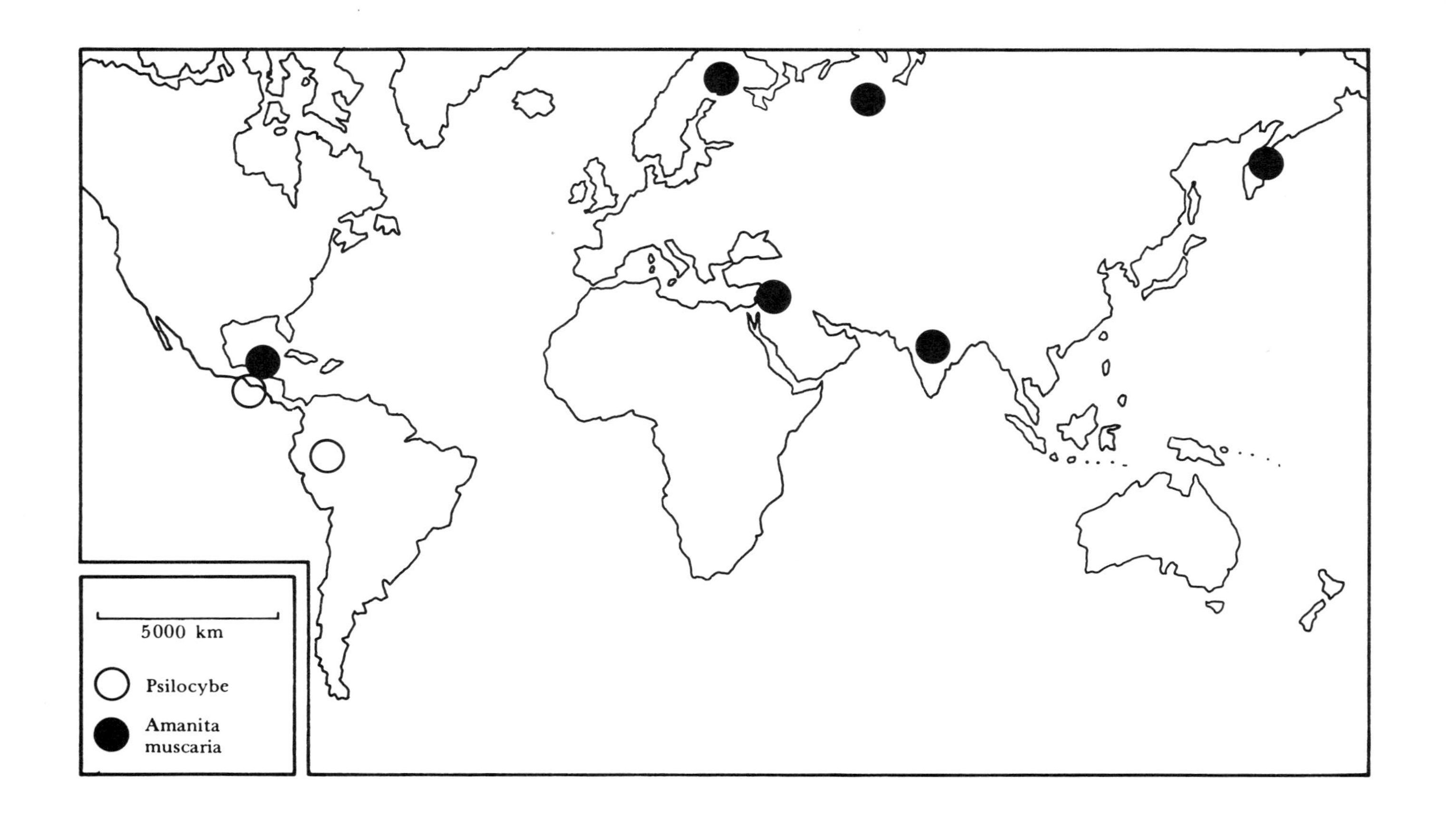
5000 km
Psilocybe
Amanita
muscaria

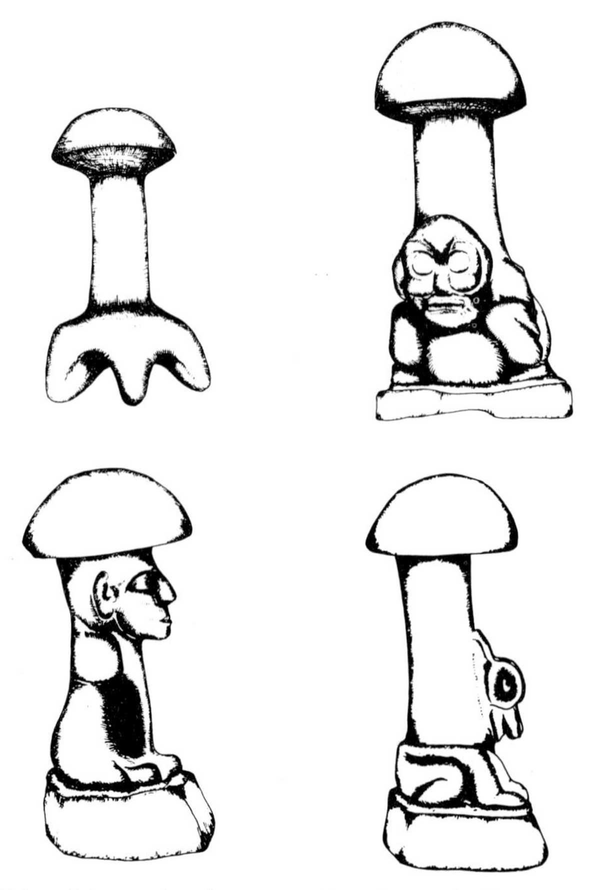

Fig. 10.6 Plain and decorated mushroom stones from Guatemala. The stones range from 28–34 cm in height. (After Lowy, 1971.)

may not be described, but may be referred to in ambiguous, euphemistic terms which perhaps reflect the awe with which it was regarded. Mycological certainty exists with respect to only one such religion. In about 1500 BC a great migration of Aryan peoples took place from the north into the Indian sub-continent, and these took with them the cult of *soma*. The juice of *soma* produced a holy inebriation and was widely utilized in religious rites, but gradually its use was for some reason abandoned and its identity was finally forgotten. From this ancient Vedic religion, which was based upon *soma*, arose Hinduism. More than a thousand hymns to *soma* have come down intact, preserved within a work known as the *Rig Veda*. Many of these contain, in metaphorical form, descriptions of the sacred, hallucinogenic fungus. Without any need to suspend belief, or to indulge in complicated interpretation, these outline in a remarkably accurate way the morphological characteristics of *A. muscaria*, and no hymn contains information that contradicts this identification.

Moving on less certain ground it is possible to adduce some evidence for the past existence of Fly-Agaric cults in other parts of the world. Mushroom-shaped stone carvings, sometimes incorporating human or animal figures on their stems, have frequently been found in Central America and are relatively common in the Highlands of Guatemala (Fig. 10.6). These are usually about 30 cm in height and are portable. Their precise function, if any, is unknown but they are thought to be relics of a Mayan practice which involved the ceremonial consumption of fungi. The mushroom stones do not overtly represent any particular kind of fruit body, but a clue as to the possible identity of one of the fungi that was used is available from other sources. A few Mayan writings which have survived in a handful of bark documents, known as the Maya Codices, contain religious or astrological paintings, some of which depict figures offering or holding a mushroom-shaped object. These objects have been variously interpreted as being calendar wheels, rattles, or fans, but they also very closely resemble basidiocarps of *A. muscaria*. The warty pileus, the ring, and even the gills and volva are quite

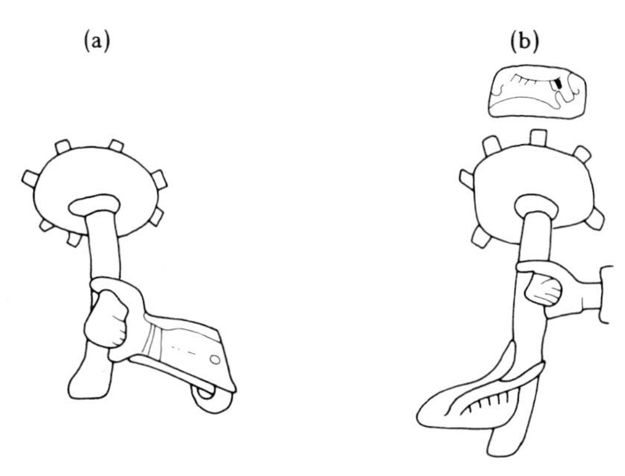

Fig. 10.7 Representations of fungi in Maya Codices: (a) part of page Lld of the Madrid Codex depicting a hand holding a basidiomycetous fruit body; (b) part of page XCVb of the Madrid Codex with a similar fruit body. The ring on the stem and the scales on the cap are clear and in (b) there appears to be a basal volva. Above the fruit body in (b) the glyph *cimi* (death) has been drawn. The mushroom may be *Amanita muscaria*. (After Lowy, 1972.)

clearly shown in different paintings (Fig. 10.7). In one painting a single glyph, *cimi*—a death symbol—has been drawn close to and above the cap, and may be referring directly to the toxic properties of the fruit body. *Amanita muscaria* is common throughout the Guatemalan sierra and this, together with the known archaeological evidence, very strongly suggests that this fungus played a powerful part in Mayan religious life.

Without very good reason, the Fly-Agaric has also been associated with the Viking berserkers and with cults in Ancient Greece. It has, for example, been suggested that the golden fleece which Jason sought was not a literal fleece but the fruit body of *A. muscaria*. More seriously, and most controversially, it has been proposed that Christianity has its roots in a widespread Fly-Agaric cult. The arguments for this view have been made largely in the field of comparative philology, and complex reasoning has been used to deduce that many Old Testament names and stories are euphemisms for *A. muscaria* or its properties. It has been suggested that by the time that the books of the New Testament were being written the mushroom cult was fragmented and under some pressure from the Jewish and Roman authorities. Much of the New Testament was therefore composed in a cryptographic form to convey to the scattered pockets of believers the doctrines and incantatory names connected with ceremonial use of the fungus, and in this way to deceive authority with a seemingly innocuous series of stories. At the centre of these stories was the Jesus 'myth' which embodied a fundamental, complex euphemism for the expanding, sacred fruit body.

This is not the place to comment on the correctness of these arguments in the context of biblical scholarship, but it is appropriate to assess the strength of the strictly mycological evidence. First, there are no pictures or cult objects extant which directly indicate the existence, nature or extent of the putative Fly-Agaric cult, or for that matter of any mushroom cult in the Middle East. Secondly, unlike the precise Vedic hymns, the names and biblical passages that are taken to be epithets for, or descriptions of, the fungus are extremely vague, and they certainly do not constitute, either singly or together, a very accurate description of *A. muscaria*. It is possible to read into these inchoate metaphors—if that is in fact what they truly are—almost anything that takes the fancy. The following, not untypical, example is from Revelations and refers to a figure seen in a vision '. . . one like a son of man clothed in a long robe, with a golden girdle round his head and his hair was white as white wool, white as snow, his eyes were like a flame of fire, his feet were like burnished bronze, ...'

The factors responsible for the narcotic and other effects of *A. muscaria* have been studied for over a century, but the exact nature of all the compounds concerned have still not been fully determined. Two alkaloids, muscarine and bufotenine, are present in trace amounts but have no hallucinogenic properties (Fig. 10.8 and 10.9). They do, however, produce other unpleasant effects such as perspiration, lachrymation, and salivation; together with nausea, vomiting and diarrhoea. In rare instances there may be death from respiratory or cardiac failure. Central nervous system activity, and thus the induction of hallucinations, seems to be due primarily to muscimol, a cyclic acid, acting together with two cyclic amino acids, muscazone and ibotenic acid (Fig. 10.8). Other as yet uncharacterized compounds may also play a part.

Mention has been made of the variability of the effect of ingesting *A. muscaria*, and this may be due in part to variations in the intoxication potential of different fruit bodies, or to seasonal variations in their biochemical composition. Ibotenic acid is a precursor of muscazone and the latter compound is not very potent. Differences in intoxication potential might therefore depend on the ratio between these two amino acids. Investigators of *Amanita*-based cults have carried out experiments on the effects of ingesting fruit body tissues using themselves as test subjects. Results have been somewhat disappointing, since many of the unpleasant physical effects of the toxins have been experienced without the compensation of inspiring visions. The reasons for this are not clear. Variation in intoxication potential might indeed be the major cause but is probably not responsible in all cases. It is more likely that the potential of the subject to respond to the narcotic compounds is of prime importance. Those compounds found in *A. muscaria* are biochemically different from those present in other hallucinogenic fungi, and the brain may have to be predisposed in some way in order to be stimulated to the necessary extent for hallucination to take place. The mind of modern man, relatively unburdened by superstition and supported by a degree

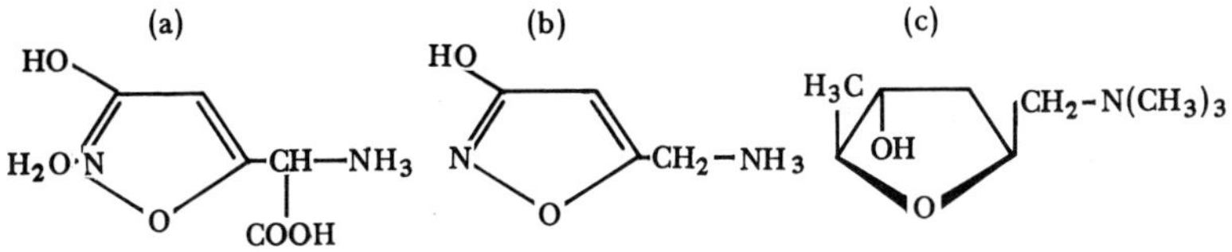

Fig. 10.8 Toxic compounds from *Amanita muscaria*: (a) ibotenic acid; (b) muscimol; (c) muscarine. A further compound, muscazone, resembles ibotenic acid but has a nitrogen atom in the ring in place of the oxygen atom.

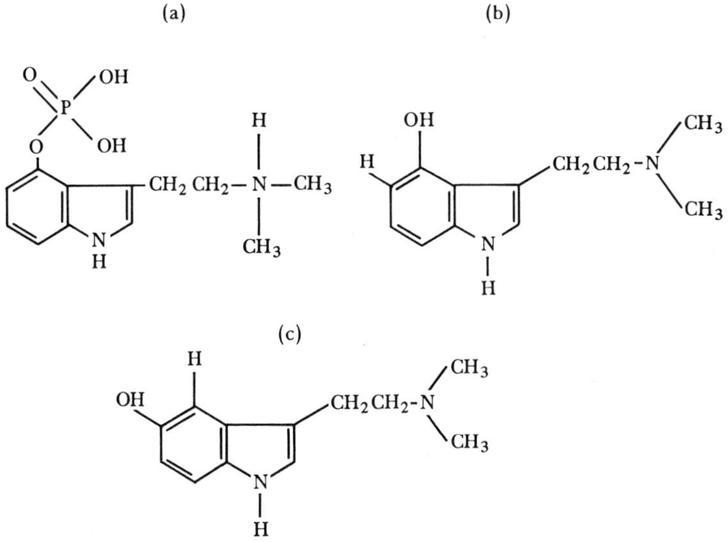

Fig. 10.9 Hallucinogenic and toxic compounds from Basidiomycete fruit bodies: (a) psilocybin; (b) psilocin. Both compounds are hallucinogens found in *Psilocybe* species. (c) Bufotenine, a toxin from *Amanita muscaria*.

of scepticism, might not be sufficiently predisposed and so could not be expected to respond when stimulated by the hallucinogens of the Fly-Agaric.

Psilocybe mexicana is a hallucinogenic mushroom that is used in parts of Mexico where it is gathered, together with other species of the same genus, either for personal consumption or for sale in the local markets. It is interesting to note that another *Psilocybe* species, *P. yungensis* may have been employed in the past as an intoxicant by some tribes of Amazonian Peru, but this practice has not persisted to the present day. *Psilocybe* species are used either for directly medical or for religious purposes. In those regions where they are utilized primarily as medicines their hallucinogenic properties seem to be considered incidental to their supposed healing qualities. There is abundant evidence that the contemporary, and possibly declining, use of these fungi in Mexico is a relic of their ceremonial role in Aztec religion. During and after their conquest of Mexico in the sixteenth century, the Spaniards frequently noted the use or narcotic mushrooms by the Indians. Fruit bodies were eaten and chocolate drunk before dawn and, after the ensuing period of hallucinations, the visions were discussed by the participants. Divination, prophecy and the worship of deities were all bound up in such practices, the details of which are now lost. The collective name for the species used was *teonanacatl*, which has been translated as 'flesh of the gods' but which literally means 'dangerous mushroom'.

Many, if not all, *Psilocybe* species contain psilocybin and psilocin, and it is these compounds which are responsible for the effects of eating the mushrooms (Fig. 10.9). Psilocybin is also produced in the mycelium of *Psilocybe* species when these are grown in pure culture under controlled conditions in liquid media. Psilocybin is chemically related to lysergic acid (Fig. 8.6) from which other and more potent psychoactive drugs can be derived, including lysergic acid diethylamide, familiarly known as LSD. Much adverse publicity has been given to the use of hallucinogenic drugs but such compounds can, if used ethically, be beneficial. For instance, it may be possible to use them in the chemotherapeutic treatment of alcoholism or to employ them as an adjunct in the psychotherapy of psychotic disturbances. Such compounds are looked upon as being non-addictive but are nevertheless not harmless, particularly if taken regularly over long periods. It has been noted that in Mexico professional divinators who make use of *Psilocybe* species age rapidly and may become senile before middle life. It has also been established that LSD can cause chromosomal damage in regular users.

Those hallucinogenic fungi so far discovered probably represent only a fraction of the total that exist, and it is almost certain that such species are not restricted to any particular group of geographical locations but that they are widespread throughout the world. In addition, the relatively uncomplicated chemical structure of psilocybin and psilocin would seem to preclude their being confined to a specific group of closely-related fungi. This raises the question as to why religious practices based on fungi arose in distinct and well-separated regions. For instance, *Psilocybe semilanceata*, 'liberty caps', is widespread in Europe and other north-temperate zones, and the fruit bodies contain sufficient psilocybin to have a hallucinogenic effect if they are eaten in even small quantities. This fungus is common, occurs in large numbers, is not poisonous, and is distinctive enough not to be confused with other species—yet, apparently, it has never been used even as a casual inebriant. Such a situation is difficult to explain, but the use of 'magic' mushrooms presumably arose from periods of experiment in which different groups of plants or fungi were tested for their edibility or for their curative properties. Whether hallucinations were accepted as a tolerable or desirable effect would possibly have depended on the degree of sophistication or of the religious orientation of the peoples involved. Rejection of hallucinogenic mushrooms, with the concomitant labelling of them as dangerous or evil, would have effectively excluded them from further use as long as there was a persistent and unfavourable folk memory of them.

Epilogue

Fungi and the future

'You are at the edge of promises and prophesies.'
Oliver Cromwell, addressing the Barebones Parliament, July 1653.

Scientific interest in fungi, as reflected by the number of published papers and books concerned with them, has been steadily growing over the past 40 years. This interest is by no means purely an academic one, and there has been a parallel increase in the effort and resources devoted to applied aspects of mycology. Fungi will always impinge on man in a number of positive and negative ways, although the manner in which they do so may alter as the apparatus of technology and the ecological impact of man gradually change with time.

On the positive side, fungi will continue to provide commercially useful enzymes and medically valuable antibiotics, but their products might also be applied in new ways or their use, at present rather restricted, may become more common. For example, fungi that produce haemolytic enzymes may become the source of anticoagulant preparations for the treatment of various kinds of thrombosis. The use of known psychoactive fungal metabolites for the treatment of some states of mental disturbance may broaden and there is a strong possibility that compounds as yet unknown will be discovered. Within Western countries there are signs that the incidence of disease attributable to stress is rapidly increasing. Psychoactive drugs used at low dosage levels and with careful controls may provide treatment for the alleviation of some stress conditions. It is certain that fungi will eventually play a major role in large-scale food production, in particular the manufacture of food proteins. It is also probable that methods of production will be linked to methods of disposal of solid and liquid pollutants, which will reduce the cost of the food protein and at the same time reduce the amount of ecologically undesirable industrial and domestic waste.

On the negative side, it is unlikely that there will be any major diminution in the importance of fungi in causing plant diseases. Despite the development of efficient fungicides and the breeding of disease-resistant crop varieties, constant vigilance and research will still be required. On numerous occasions some species of fungi have demonstrated their genetic versatility to overcome host resistance and an ability to appear suddenly and catastrophically among crops not previously exposed to them. It is arguable that the moulding of a new variety of crop plant inevitably results, with time, in the evolution of a strain of a fungus that is able to attack it. Similarly, the importance of fungi in diseases of animals, including man, is likely to increase: at the moment, aspects of animal-parasitic fungi provide one of the great growth areas in mycology.

Increasingly the fungi have come to be looked upon with favour as tools for fundamental biological research. They are easy to maintain in pure culture, have relatively simple nutrient requirements and can be grown rapidly under controlled conditions. To the molecular biologist, biochemist and physiologist these are valuable properties for an organism to possess if it is to be readily used in laboratory experiments. Much of our knowledge of the basic metabolic processes that take place within living cells is based upon the study of fungi.

Finally, the recent landing of the Viking spacecraft on Mars has provided evidence for the existence of conditions which could support microbial life and has, perhaps, detected that life itself. It is not beyond the realms of possibility that fungi, or fungus-like life forms, dwell there, and that when samples of Martian soil are, in the not too distant future, brought back to Earth, the science of exomycology will be born. It can fairly be said that the New Kingdom has now come into its own.

Bibliography

Since this is not a textbook the list of references given below is by no means exhaustive but contains selected research papers, review articles and books that present easily-assimilable information on particular topics.

1. The New Kingdom

Martin, G. W. (1955). Are fungi plants? *Mycologia*, **47,** 779–92.

Whittaker, R. H. (1969). New concepts of kingdoms of organisms. *Science*, **163,** 150–60.

2. Nutrients and niches

Crisan, E. V. (1973). Current concepts of thermophilism and the thermophilic fungi. *Mycologia*, **65,** 1171–98.

Hendey, N. I. (1964). Some observations on *Cladosporium resinae* as a fuel contaminant and its possible role in the corrosion of aluminium alloy fuel tanks. *Transactions of the British Mycological Society*, **47,** 467–75.

Hudson, H. J. (1972). *Fungal Saprophytism.* Edward Arnold, London.

Parberry, D. G. (1969). The natural occurrence of *Cladosporium resinae. Transactions of the British Mycological Society*, **53,** 15–23.

3. Crop diseases and natural catastrophes

Austin Bourke, P. M. (1964). Emergence of potato blight, 1843–6. *Nature*, **203,** 805–8.

Brasier, C. M. and Gibbs, J. N. (1973). Origin of Dutch elm disease epidemic in Britain. *Nature*, **242,** 607–9.

Carefoot, G. L. and Sprott, E. R. (1967). *Famine on the Wind: man's battle against plant disease*. Rand McNally, New York.

Corbaz, R. (1964). Evolution de l'épidémie de mildiou de tabac (*Peronospora tabacina* Adam). *Phytopathologische Zeitschrift*, **51,** 191–2.

Gibbs, J. N. and Howell, R. S. (1974). Dutch elm disease survey 1972–3. *Forestry Commission Forest Record*, No. 100. HMSO, London.

Large, E. C. (1940). *The Advance of the Fungi*. Jonathan Cape, London.

Newhook, F. J. and Podger, F. E. (1972). The role of *Phytophthora cinnamomi* in Australian and New Zealand forests. *Annual Review of Phytopathology*, **10,** 299–326.

Schieber, E. (1972). Economic impact of coffee rust in Latin America. *Annual Review of Phytopathology*, **10,** 491–510.

Tatum, L. A. (1971). The southern corn leaf blight epidemic. *Science*, **171,** 1113–1116.

Ullstrup, A. J. (1972). The impacts of the southern corn leaf blight epidemics of 1970–1. *Annual Review of Phytopathology*, **10,** 37–50.

Weste, G. (1974). *Phytophthora cinnamomi*—the cause of severe disease in certain native communities in Victoria. *Australian Journal of Botany*, **22,** 1–8.

Woodham-Smith, C. (1962). *The Great Hunger, Ireland 1845–1849*. Hamish Hamilton, London.

4. Symbioses and dual organisms

Gilbert, O. L. (1970). Further studies on the effect of sulphur dioxide on lichens and bryophytes. *New Phytologist*, **69,** 605–27.

Gilbert, O. L. (1970). A biological scale for the estimation of sulphur dioxide. *New Phytologist*, **69,** 629–34.

Hale, M. E. (1974). *The Biology of Lichens*. Edward Arnold, London.

Harley, J. L. (1969). *The Biology of Mycorrhiza*. Leonard Hill, London.

Mosse, B. (1973). Advances in the study of vesicular-arbuscular mycorrhiza. *Annual Review of Phytopathology*, **11,** 171–96.

Stribley, D. P. and Read, D. J. (1974). The biology of mycorrhiza in the Ericaceae. IV. The effect of mycorrhizal infection on uptake of ^{15}N from labelled soil by *Vaccinium macrocarpon* Ait. *New Phytologist*, **73,** 1149–55.

5. Ailments of man and his livestock

Emmons, C. W. (1960). The Jekyll–Hydes of mycology. *Mycologia*, **52,** 669–680.

Emmons, C. W., Binford, C. H. and Utz, J. P. (1970). *Medical Mycology*. Lea and Febiger, Philadelphia.

Hammerman, K. J., Powell, K. E. and Tosh, F. C. (1974). The incidence of hospitalized cases of systemic mycotic infections. *Sabouraudia*, **12,** 33–45.

Sindermann, C. J. (1970). *Principal diseases of marine fish and shellfish*. Academic Press, London and New York.

Stuart, M. R. and Fuller, H. T. (1968). Mycological aspects of diseased atlantic salmon. *Nature*, **217,** 90–2.

Unestam, T. (1973). Fungal diseases of crustacea. *Review of Medical and Veterinary Mycology*, **8,** 1–20.

6. Allies against pests

Burgess, H. D. and Hussey, N. W. (Eds.) (1971). *Microbial Control of Insects and Mites*. Academic Press, London.

Conway, K. E. (1976). Evaluation of *Cercospora rodmanii* as a biological control of water hyacinths. *Phytopathology*, **66,** 914–17.

Cullen, J. M., Kable, P. F. and Catt, M. (1973). Epidemic spread of a rust imported for biological control. *Nature*, **244,** 462–4.

Rishbeth, J. (1963). Stump protection against *Fomes annosus*. III. Inoculation with *Peniophora gigantea*. *Annals of Applied Biology*, **52,** 63–77.

Rishbeth, J. (1976). Chemical treatment and inoculation of hardwood stumps for control of *Armillaria mellea*. *Annals of Applied Biology*, **82,** 57–70.

Wapshere, A. J. (1974). A strategy for evaluating the safety of organisms for biological weed control. *Annals of Applied Biology*, **77,** 201–11.

Zettler, F. W. and Freeman, T. E. (1972). Plant pathogens as biocontrols of aquatic weeds. *Annual Review of Phylopathology*, **10,** 455–70.

7. Secret gardens and insect hosts

Buchner, P. (1965). *Endosymbiosis of Animals with Plant Microorganisms*, Wiley, New York.

Couch, J. N. (1938). *The genus Septobasidium*. University of North Carolina Press.

Graham, K. (1967). Fungal–insect mutualism in trees and timber. *Annual Review of Entomology*, **12,** 105–26.

Francke-Grosmann, H. (1967). Endosymbiosis in wood-inhabiting insects. In *Symbiosis*, Volume II, ed. by. S. M. Henry, Academic Press, New York, pp. 141–205.

Parkin, E. A. (1942). Symbiosis and Siricid wood wasps. *Annals of Applied Biology*, **27,** 268–74.

Weber, N. A. (1972). *Gardening ants—the Attines*. American Philosophical Society, Philadelphia.

Weber, N. A. (1966). The fungus-growing ants. *Science*, **153,** 587–604.

8. Toxins

Lacey, J. (1975). Potential hazards to animals and man from microorganisms in fodders and grain. *Transactions of the British Mycological Society*, **65,** 171–84.

Litten, W. (1975). The most poisonous mushrooms. *Scientific American*, **232,** 90–101.

Spensley, P. C. (1963). Aflatoxin, the active principle in turkey X disease. *Endeavour*, **22,** 75–9.

Wright, D. E. (1968). Toxins produced by fungi. *Annual Review of Microbiology*, **22,** 269–82.

9. Industrial protein and modified foods

Brian, P. W. (1972). The economic value of fungi. *Transactions of the British Mycological Society*, **58,** 359–75.

Hayes, W. A. (1969). Mushrooms, microbes and malnutrition. *New Scientist*, **44,** 450–2.

Hayes, W. A. (1974). Mushroom cultivation—prospects and developments. *Process Biochemistry*, **9,** 21–8.

Hering, T. F. (1966). The terricolous higher fungi of four Lake District woodlands. *Transactions of the British Mycological Society*, **49,** 369–83.

Hesseltine, C. W. (1965). A millenium of fungi, food and fermentation. *Mycologia*, **57,** 149–97.

Humphrey, A. E. (1970). Microbial protein from petroleum. *Process Biochemistry*, **5,** 19–22.

Reade, A. E. and Gregory, K. F. (1975). High temperature production of protein-enriched feed from cassava by fungi. *Applied Microbiology*, **30,** 897–904.

Shacklady, C. A. (1974). SCP from hydrocarbons as animal feed ingredients. *Process Biochemistry*, **9,** 9–14.

Shedwood, M. (1974). Single-cell protein comes of age. *New Scientist*, **64,** 634–639.

Vilenchich, R. and Akhtar, W. (1971). Microbiological synthesis of proteins. *Process Biochemistry* **6,** 41–4.

10. Magic mushrooms and hallucinogenic drugs

Allegro, J. M. (1970). *The Sacred Mushroom and the Cross.* Hodder and Stoughton, London.

Borhegyi, S. (1961). Miniature mushroom stones from Guatemala. *American Antiquity*, **26,** 498–504.

Borhegyi, S. (1963). Pre-Columbian pottery mushrooms from Mesoamerica. *American Antiquity*, **28,** 328–38.

Lowy, B. (1971). New records of mushrooms stones from Guatemala. *Mycologia*, **63,** 983–93.

Lowy, B. (1972). Mushroom symbolism in Maya codices. *Mycologia*, **64,** 816–821.

McElroy, W. D. and Seliger, H. H. (1962). Biological luminescence. *Scientific American*, **207,** 76–89.

Mantle, P. G. and Wright, E. S. (1969). Occurrence of psilocybin in the sporophores of *Psilocybe semilanceata*. *Transactions of the British Mycological Society*, **53,** 302–4.

Schultes, R. E. (1963). Hallocinogens of plant origin. *Science*, **163,** 245–54.

Wasson, R. G. (1971). *Soma—Divine Mushroom of Immortality.* Harcourt Brace, Jovanovich, New York.

Index